W9-CRB-422

Crop Biotechnology

ACS SYMPOSIUM SERIES **829**

Crop Biotechnology

K. Rajasekaran, Editor
U.S. Department of Agriculture

T. J. Jacks, Editor
U.S. Department of Agriculture

J. W. Finley, Editor
Kraft Foods

American Chemical Society, Washington, DC

Library of Congress Cataloging-in-Publication Data

Crop biotechnology / K. Rajasekaran, editor, T. J. Jacks, editor, J. W. Finley, editor.

p. cm.—(ACS symposium series ; 829)

Includes bibliographical references and index.

ISBN 0–8412–3766–2

1. Plant biotechnology—Congresses. 2. Crops—Genetic engineering—Congressess.

I. Rajasekaran, K., 1952-. II. Jacks, T. J. (Thomas J.), 1938-III. Finley, John W., 1942- IV. American Chemical Society. Meeting (219th: 2000 : San Francisco, Calif.). V. Series.

SB106.B56 C76   2002
631.5′233—dc21   2002018690

The paper used in this publication meets the minimum requirements of American National Standard for Information Sciences—Permanence of Paper for Printed Library Materials, ANSI Z39.48–1984.

PRINTED IN THE UNITED STATES OF AMERICA

Foreword

The ACS Symposium Series was first published in 1974 to provide a mechanism for publishing symposia quickly in book form. The purpose of the series is to publish timely, comprehensive books developed from ACS sponsored symposia based on current scientific research. Occasion-ally, books are developed from symposia sponsored by other organizations when the topic is of keen interest to the chemistry audience.

Before agreeing to publish a book, the proposed table of contents is reviewed for appropriate and comprehensive coverage and for interest to the audience. Some papers may be excluded to better focus the book; others may be added to provide comprehensiveness. When appropriate, overview or introductory chapters are added. Drafts of chapters are peer-reviewed prior to final acceptance or rejection, and manuscripts are prepared in camera-ready format.

As a rule, only original research papers and original review papers are included in the volumes. Verbatim reproductions of previously published papers are not accepted.

ACS Books Department

Contents

Indexes

Preface

The world population is increasing by 80 million people per year according to the Food and Agriculture Organization of the United Nations. The greatest increase is occurring in developing countries where food requirements are expected to double in 25 years. And, of course, concomitant with the population increase is the greater proportion that will suffer malnutrition and even starvation. These sufferings can largely be offset by increased food production, which can be accomplished in three ways: by increasing the use of chemical fertilizers and pesticides, by increasing the area of land under tillage, and by obtaining higher yields from crop plants. Chemicals are already in heavy use in farming areas of developed countries, so using more won't gain much. Most arable land is already under tillage. That leaves increasing food production in crops on hand.

Crop productivity and distribution are limited principally by biotic stresses and growth environments. Insect feeding and pathogen infection, which comprise major biotic stresses, are fought with chemical pesticides. Growth environments are modified by irrigation, by chemical fertilizers, and by tools such as smudge pots and fans to prevent airborne frost. A far more intelligent method is to modify the plant per se to overcome stress and to fit the environment. In this manner over-dependence on chemical fertilizers and pesticides that are poisoning us and our fragile environment is reduced, food yields are increased in tandem with the burgeoning population, and nutrient contents and safety of foods are improved for our health and general wellness. Combined with improved pesticides and fertilizers, the development of high yielding varieties of grains and rice by conventional breeding jump-started the *Green Revolution*. In the current *gene revolution,* plant modification is accomplished by genetic engineering, defined as the introduction of foreign genes by any means other than those employed by conventional breeding. With this technology, plants have been engineered for cold hardiness, herbicide tolerance, salt tolerance, water stress resistance, pathogen resistance, and insect resistance among others. In other words, rather than artificially modifying the environment to suit plants, a goal of genetic engineering is to modify plants to fit the environment. Of course, corollary studies on transgenic plants and resultant food products are conducted to ensure environmental and consumer safety.

Genetic engineering also provides plants with the properties of *biological factories* for the production of nonfoods. For instance, plants can be engineered to synthesize both botanical and animal natural products of industrial and pharmaceutical/medical importance. Through microbial genetic engineering, biotechnology companies have developed several human health products (e.g., insulin, blood clotting factor, growth hormone, cancer treatment compounds). Recently, crop plants rather than microorganisms are being engineered to produce human health products. Indeed, the epithet *molecular pharming* has been coined to describe this technology.

This book is a product of a 3-day symposium held during the 219th American Chemical Society (ACS) national meeting in San Francisco, California in 2000 as part of "A Celebration of Chemistry in the 21st Century". The purpose of the symposium was to provide a forum for conveners specializing in the technologies and problems involved in plant transformation. These specialists were involved in a variety of scientific interests, with plant biotechnology as the common thread, and were sometimes working in isolation with limited personal interaction. Publication of the resultant peer-reviewed papers provides a source of information for specialists, for professionals and students of related and unrelated disciplines, and for the concerned public regarding the area of genetic engineering of crop plants. Thus, the book meets the objective of providing a full comprehension of genetically engineered crops and demonstrates the usefulness of this powerful technology. The chapters deal with understanding the issues in the challenging task of both creating genetically engineered plants with selective functions as well as the impact of genetically engineered plants on nutrition, environment, and economy. The reader will derive a basic understanding of some of the current approaches being employed in applying biotechnology research to crop production and utilization. For those less familiar with genetic engineering, a short introductory chapter provides a general overview.

As the philosopher Arthur Schopenhauer (1788–1860) noted, "Every truth passes through three stages before it is recognized. In the first stage, it is ridiculed. In the second, it is opposed. In the third, it is regarded as self-evident". History dictates that new technologies also proceed through these steps. The firm basis of genetic engineering on scientific principles has rendered it above ridicule but not yet beyond opposition.

Acknowledgments

The organizers of the symposium gratefully acknowledge partial funding from the ACS Division of Agricultural and Food Chemistry and from the USDA National Research Initiative (NRI) Competitive Grants Program Award (No. 00–35300–8852). We also appreciate valuable assistance from Dr. Mike Morello, the ACS Division of Agricultural and Food Chemistry, and the staff of the ACS Books Department.

K. Rajasekaran
Research Biologist
Southern Regional Research Center
U.S. Department of Agriculture
1100 Robert E. Lee Boulevard
New Orleans, LA 70124
Phone: 504–286–4482
Fax: 504–286–4419
Email: krajah@srrc.ars.usda.gov

T. J. Jacks
Research Biologist
Southern Regional Research Center
U.S. Department of Agriculture
1100 Robert E. Lee Boulevard
New Orleans, LA 70124
Phone: 504–286–4380
Fax: 504–286–4419
Email: tjacks@srrc.ars.usda.gov

J. W. Finley
Kraft Fellow
Kraft Foods
801 Waukegan Road
Glenview, IL 60025
Phone: 847–646–7321
Fax: 847–646–3668
Email: jfinley@kraft.com

Chapter 1

Overview of Crop Biotechnology

K. Rajasekaran[1], T. J. Jacks[1], and J. W. Finley[2]

[1]Southern Regional Research Center, Agricultural Research Service,
U.S. Department of Agriculture, 1100 Robert E. Lee Boulevard,
New Orleans, LA 70124
[2]Kraft Foods, 801 Waukegan Road, Glenview, IL 60025

Scope

The influence of crop biotechnology on outcomes of agricultural practices and economics is readily evidenced by the escalating acreage of genetically engineered crops, all occurring in a relatively short time span. Until the mid 1990s, virtually no acreage was planted with commercial genetically modified (GM) crops worldwide. But by 2001 transgenic plants globally comprised 46% of the soybean crop (equivalent to 82 million acres), 20% of the cotton crop (17 million acres), 11% of the canola crop (7 million acres) and 7% of the corn crop (24 million acres). Combining areas for these four principal crops shows that 19% of the total acreage under tillage supports GM crops. Amazingly, in just six years, the total area planted with GM crops increased by more than 30-fold, from slightly more than 4 million acres in 1996 to 130 million in 2001 *(1)*! Although the amount of farmland acreage of GM crops in developed countries has become stable, this is not the case in developing countries. Additional biotechnological developments of staple grain crops such as rice and wheat will put even more acreage into GM crop production. Furthermore, with a nationwide research staff of about 2000 members studying crop biotechnology supported by a budget eclipsing that in 1999 of $112 million, China will add substantially to global GM crop acreage in the future *(2)*. Indeed, over 50 plant species and more than 120 functional genes are currently being employed in the development of GM plants, making China a global leader in crop biotechnology.

1

History

Mankind's ability to control and even manipulate the properties of living organisms for his advantage is based on the principles of heredity discovered by Mendel, published in 1866. With controlled pollination and prudent statistical analysis, his observations concerning the inheritance of phenotypic traits in plants led to the formulation of the laws of segregation and independent assortment. His principles of factorial inheritance with the quantitative investigation of single characters provided the basis for modern genetics. Consequentially, identification and transference of desirable traits in plants through selection and breeding brought mankind improved food crops, of which hybrid corn developed in the 1920s was the first major achievement and is probably the most widely recognized. The impact of classical breeding is amply evident in the "Green Revolution" of the 1960s in which semi-dwarf genes were transferred into wheat and rice to decrease yield losses from lodging and to increase crop productivity. Consequently, about one billion people were saved from starvation, and countries on the brink of famine were even able to become grain exporters years later. Crop improvement through conventional breeding, however, is a slow and tedious process. For example, it takes up to seven generations to release a new variety of cotton, and tree crop breeders might never see results in their lifetime! The current "Gene Revolution" based on techniques of genetic engineering allows scientists to rapidly identify and disseminate genetic material (DNA), aptly called genetic "blue prints" of living organisms, among widely diverse organisms. Using genetic engineering technology, scientists are able to transfer genes among plants and even bridge supposedly "natural barriers," such as those between widely divergent organisms, e.g. gene transference from viruses, bacteria, or even animals, to plants. Furthermore, using modern biotechnological methods, researchers are able to quickly identify and evaluate their results to determine whether the gene manipulations were successful in the first generation of transgenic plants.

During this period, microbiologists discovered that bacteria, which usually divide by binary fission, are able to obtain new phenotypic traits not only by conventional reproductive mating (conjugation) but also by incorporating pieces of foreign DNA from their environment into their plasmid DNA, which replicates independently of the host bacterium's chromosomes. This incorporation (transformation) was found to occur in nature and could be accomplished by physicochemical manipulations in the laboratory. Thus, microorganisms were "genetically engineered" and scientists subsequently ascertained that this occurrence could happen in higher organisms such as plants.

The actual roots of genetic engineering can be traced to the early 1970s with the research accomplishments of Paul Berg, Stan Cohen and Herb Boyer. They inserted fragments of DNA prepared with "molecular scissors" called a restriction enzyme into circular plasmid DNA that had also been cut with the enzyme. Using segments of DNA from different organisms in combination with DNA ligase, they produced new DNA, called "rDNA" for recombinant DNA. Transformation of bacteria with rDNA yielded the first genetically engineered or genetically modified organisms (GMO). DNA from one species was successfully incorporated into the genome of a different species!

Background

For the purposes of introducing the subject of crop biotechnology to the general reader and reviewing it for others, techniques of plant transformation are briefly described here. If a foreign protein such as an enzyme is to be introduced into a plant that lacks it, the corresponding DNA (gene) must first be identified by sequencing techniques and then prepared for insertion into the target cell DNA. The first step is usually achieved by using short segments of nucleic acids (primers) to fish out similar (homologous) genes from DNA isolated from the donor as the template. Only the gene of interest, for which primers were provided, is copied by the polymerase chain reaction (PCR). Alternatively, if the DNA sequence of the gene has been deposited in a readily available "gene bank," the gene can be synthesized directly. Isolated genes are then introduced into bacterial plasmid- or viral- "vector" molecules for further multiplication and/or subsequent transference of DNA fragments to target plants. The actual plant transformation process technically consists of two-steps: 1) insertion of the foreign gene into the target cell genome using the methods described below, and 2) regeneration of a whole plant from the resultant transformed cell. Depending on the plant species, the second step can often be the more difficult step.

The most commonly used method for gene insertion exploits the natural infection of plants by pathogenic *Agrobacteria*. Tumor- or root-inducing strains of *Agrobacterium tumefaciens* or *A. rhizogenes* infect wounded cells of many dicotyledonous plants resulting in development of tumors or roots, respectively, within a week of infection. Working independently, the requirement of Ti or tumor-inducing plasmid for the virulence of *A. tumefaciens* was first reported in 1975 by two different groups: Schell, Schilperoort, van Montague and their colleagues of the Gent-Leiden team *(3)*, and Chilton and her colleagues of the Seattle team *(4)*. Normally upon infection, the bacterium modifies plant cell growth by inserting a piece of its plasmid DNA, called "T-DNA" for transfer DNA, into the nuclear DNA of the host cell. In the case with *A. tumefaciens*, the ultimate result, orchestrated by T-DNA insertion, is a disease called "crown gall" due to production of growth hormones. The bacterially transformed cells in the gall tissue supply the bacteria with required nitrogenous nutrients. This relationship can be viewed as symbiosis between bacteria and plant cells whereby plants provide a site for the bacteria and the bacteria, in return, consume plant waste products. Through processes of molecular biology, scientists eliminated the oncogenic activity of *Agrobacterium* but its ability to insert DNA into plant cells was retained. The foreign gene of interest, after being slightly modified to appear as a part of T-DNA to the bacterial machinery, is inserted into the plant cell DNA via non-oncogenic *Agrobacterium*-infection to produce transformed cells but without the attendant disease.

In the particle bombardment method, also known as "biolistics," microscopic metal particles coated with DNA are literally shot into cells by physical force, as originally described by Klein and his colleagues *(5)*. Several variations of particle bombardment have been developed and applicability of these methods for plant transformation has been reviewed *(6)*. Insect resistant corn (Bt-corn) was produced by this technique.

Plant cell transformation by inducing DNA uptake physically, such as by electroporation, or chemically, such as with polyethylene glycol, would probably be the methods of choice for gene transfer because biological vectors aren't required *(7)*. However, prerequisite steps such as isolation and culture of protoplasts (naked cells without walls) are difficult procedures for many plant species. A second physical transformation process successful with a few crop species is the use of silicon whiskers as DNA carriers *(8)*.

In addition to transforming cell nuclear DNA, plant scientists have also developed a method for transforming DNA of cell organelles, for instance that of the chloroplast genome. Transformations of chloroplast genomes not only yield high levels of gene expression, but also are based on maternal inheritance, a property that mitigates concerns over dispersal of transgenic pollen. Additional benefits of chloroplast transformation are reviewed by Maliga *(9)* and Daniell et al. *(10)*. Such research fortifies the fact that scientists constantly pursue new processes to address genuine concerns regarding genetically modified crops.

Other foreign genes, called "marker" or "reporter" genes, can be linked to the gene of interest for simultaneous insertion so that the success of transformation can be identified and evaluated. Products of diagnostic marker genes, e.g. the glucuronidase (*gus*) or luciferase (*luc*) gene, are easily observed and their detection in target cells by simple laboratory techniques indicates a high probability that the gene of interest is also present. Products of selectable marker genes confer growth advantages to successfully transformed cells. For instance, if a gene for antibiotic resistance is linked to the gene of interest, then only cells transformed with the resistance gene will survive when exposed to toxic levels of the antibiotic. As above with the diagnostic marker gene, chances are very good that the viable, proliferating cells also contain the gene of interest.

Once the transformed cells are identified, scientists can then regenerate fully fertile plants from them by growing the cells on a series of regimented artificial media supplemented with nutrients and plant hormones. Generally these plants are similar to their donor parental varieties but with the addition of the improved trait such as insect resistance, herbicide-tolerance, disease resistance or another value-added trait. After multiplication of seed supply, the transgenic varieties can be released for cultivation or used as donor parents to improve existing commercial varieties by conventional breeding.

Thus, the marriage of two powerful technologies, recombinant DNA technology and plant tissue culture, has given rise to the science of crop biotechnology to produce agronomically superior varieties.

Concerns

The public acceptance and reluctance regarding food and food ingredients derived from GM crops are complicated matters that have unfortunately evolved mainly into human health and safety issues that are beyond the scope of this book. Furthermore, a discourse on the pros and cons of crop biotechnology research and products related to these issues is also not an objective of this chapter. However, two poignant and relevant observations about acceptance of GM foods by the global population are readily evident. When people are hungry, the composition of food is irrelevant. Any food is adequate and GMO food that increases both quantity and quality of nutrients is superior. Acceptance is not a concern in developing countries where most of the population is food producers. In these countries, food consumers are also food producers. Rather, acceptance is a concern in food-rich countries where less than 5% of food consumers are food producers. When choices of food outweigh dependencies on food, members of food-rich societies influence and even dictate policies regarding GM foods in their countries.

Opponents to crop biotechnology have used the "precautionary principle" stating that insufficient evidence exists to conclude that consumption of transgenic crops and related food products is not harmful or unhealthy. Without proof that transgenic crops are not dangerous, opponents believe agricultural production should be banned. As pointed out above, people in food-rich countries might have the luxury to question or even ban production of GM crops, but people in developing countries don't. In ten years this population will be about 6 billion, equivalent to the whole population of the world today. Agricultural biotechnologists are diligently working to provide increased crop productivity.

In attempts to minimize any risks associated with transgenic crops, effort has been expended to identify and quantify risk possibilities. This involves determining the likelihood and consequence of errors. Risk quantification entails determination of "substantial equivalence," indicating that comparisons between GM crops and their unmodified counterparts can measure whether

6

modification has produced unwanted effects, either directly or indirectly. Comparisons involving changes in nutrient compositions, production of toxins, etc., in food products are needed to establish substantial equivalence. Comparisons can exist at several levels. For instance, the presence of identical properties of cooking oils from a GM oilseed and its unmodified counterpart doesn't indicate that other components wouldn't differ. To date, however, analyses of transgenic crops currently in production have not found any significant differences between them and their unmodified analogues with respect to harmful or unhealthy qualities *(11)*

As shown earlier, transformation of plants to yield transgenic progeny is based on scientific principles and practices, but negative opinions concerning the products of transgenic crops exist. In the U. S. three government agencies regulate various aspects of transgenic plants: the Animal and Plant Health Inspection Service of the U. S. Department of Agriculture, the Food and Drug Administration, and the Environmental Protection Agency. Since ninety percent of the money currently spent on food in the U. S. is spent on processed foods that could contain ingredients from GM crops, these agencies realize consumers have the right to information about problems and regulations concerning GM crops, and they provide it.

References

1. James, C. *Journal of the Science of Food and Agriculture* **2001**, *81*, 813-821.
2. Huang, J.; Rozelle, S.; Pray, T.; Wang, Q. *Science* **2002**, *295*, 674-677.
3. Van Larebeke, N.; Gentello, C.; Schell, J.; Schilperoort, R. A.; Herman, A. K.; Hernalsteens, J. P.; Van Montagu, M. *Nature* **1975**, *255*, 742-743.
4. Watson, B.; Currier, T. C.; Gordon, M. P.; Chilton, M.-D.; Nester, E. W. *Journal of Bacteriology* **1975**, *123*, 255-264.
5. Klein, T. M.; Wolf, E. D.; Wu, R.; Sanford, J. C. *Nature* **1987**, *327*, 70-73.
6. Christou, P. *Particle Bombardment for Genetic Engineering of Plants;* R. G. Landes Company and Academic Press: Austin, TX, 1996.
7. Vasil, I. K. *Advances in Agronomy* **1976**, *28*, 119-160.
8. Kaeppler, H. F.; Somers, D. A.; Rines, H. W.; Cockburn, A. F. *Theoretical and Applied Genetics* **1992**, *84*, 560-566.
9. Maliga, P. *Current Opinion in Plant Biology* **2002**, *5*, 164-172.
10. Daniell, H.; Khan, M. S.; Allison, L. *Trends in Plant Science* **2002**, *7*, 84-91.
11. Custers, R. Safety of Genetically Engineered Crops. http://www.vib.be/downloads/bioveiligheidseducatie/report.pdf , **2001**. Jo Bury, VIB, Belgium 159 pp.

Chapter 2

Defining Biotechnology: Increasingly Important and Increasingly Difficult

J. W. Radin and P. K. Bretting

National Program Leaders, Agricultural Research Service, U.S. Department of Agriculture, Room 4-2232, George Washington Carver Center, 5601 Sunnyside Avenue, Beltsville, MD 20705–5139

Biotechnology is in the news these days. From newspapers to television to the World Wide Web, it is difficult to avoid hearing something about agricultural biotechnology. Given the prominence of the issue, and the depth of feeling of those who are for it or against it (or are neither, but want its products to be specially labeled), the debate about biotechnology in society will not likely disappear anytime soon.

For society to decide how to manage biotechnology, first there must be a common understanding of what it is, and what the "rules" of management are intended to accomplish. Otherwise, the rules to be drawn up may miss their target. Actually, there are many biotechnologies, some of them ancient and widely used, some of them modern. Some of the scientific endeavors that are currently classified as biotechnology by scientists would not intuitively be so described by most lay people, and *vice versa*. Thus, confusion often permeates discussions of biotechnology. The bottom line is: definitions matter. Defining what is to be managed is like casting a fishing net into the sea. If the net catches all manner of unwanted fish in addition to the desired species, then it damages the ocean's resources and productivity. On the other hand, if it allows most of the desired fish to slip away, then it does not accomplish its intended purpose and might as well not be used.

The purpose of this paper is not to review the technical aspects of the various biotechnologies, nor is it to advocate whether or how to deploy them. Rather, the purpose is to convey a sense of how our current definitions and understandings of biotechnology developed; to identify which definitions might "catch unwanted fish" or let the intended ones slip away; and to project how definitions might determine the nature of imposed rules - especially rules to support labeling of the products of biotechnology.

7

What is Biotechnology?

In the 1970s, recombinant DNA technology was in its infancy. Biotechnology was often expected to generate achievements that had previously been completely infeasible, for example corn plants capable of hosting nitrogen-fixing bacteria and thereby providing their own fertilizer, like soybeans or peanuts. Because of the magnitude of its perceived potential, the plant science research community expected funding of plant biotechnology research to increase rapidly. Thus, there was substantial incentive to scientists to define biotechnology broadly, both for the association with a "glamorous" area of science and for pragmatic reasons associated with competition for limited research funding. From this understanding arose some extremely broad definitions of biotechnology. For example, the U.S. Department of Agriculture (USDA) maintains a web page about biotechnology which offers the following classical interpretation:

> Agricultural biotechnology is a collection of scientific techniques, including genetic engineering, that are used to create, improve, or modify plants, animals, and microorganisms... *(1)*

The Agricultural Research Service (ARS), the in-house research agency of USDA, classifies biotechnology research into six components: basic engineering of recombinant DNA; DNA sequencing; genomic mapping with molecular markers; monoclonal antibodies; cell fusion and chromosome transfer; and biologically-based processing. This is indeed a collection of disparate technologies, some of which have been in use for a long time. Lest this broad concept of biotechnology be considered aberrant, a recent report of the Council for Agricultural Science and Technology described biotechnology thus:

> Biotechnology refers generally to the application of a wide range of scientific techniques to the modification and improvement of plants, animals, and microorganisms that are of economic importance. Agricultural biotechnology is that area of biotechnology involving applications to agriculture. In the broadest sense, traditional biotechnology has been used for thousands of years, since the advent of the first agricultural practices, for the improvement of plants, animals, and microorganisms *(2)*.

Most other major agricultural research organizations have similar definitions.

Traditional and Modern Biotechnology

Based on these concepts, we must differentiate between traditional biotechnology, which seems largely not to be the focus of today's public discussions, and modern biotechnology, parts of which are the point of discussion. All crops are the result of breeding and selection, a traditional biotechnology technique. Traditional biotechnology products also include biologically-processed items like bread, cheese, and wine.

The modern biotechnology of interest centers upon the newfound ability to remove DNA from cells of an organism, modify it, and reinsert it into cells where it will be functional. This process is sometimes called "genetic engineering," and products therefrom have often been ingenuously termed "genetically modified organisms" (GMOs), even though all crop plants are genetically modified in one way or another. (Other modern biotechnologies, such as monoclonal antibodies or molecular markers as aids for traditional breeding and selection, are not relevant to genetic engineering and will not be treated further.) There are no theoretical species barriers to the transfer of DNA by genetic engineering; thus it is even possible to transfer genes from microbes or animals to plants, where they will change the properties of the recipient organisms. The technology is clearly related to traditional cross-breeding, as both move genes in a directed fashion. Nonetheless, it is an evolution and extension of traditional cross-breeding. The traditional breeder's available gene pool is predominantly limited to those genes in sexually-compatible organisms, whereas modern biotechnology enables some new, wider-ranging, choices.

Questions raised by these new choices have ranged from food safety to environmental safety to ethics to "natural law." Although food safety and environmental safety of new products, including those derived from modern biotechnology, are already closely examined and regulated in the United States, concern has arisen that GMOs carry with them new risks, possibly including unidentified risks, and require greater than normal regulation, including, perhaps, special labeling.

Defining Genetically Engineered Foods for Labeling Purposes

As a specific regulatory requirement, labeling, of course, requires a precise definition of what is to be labeled. In its Website, the Union of Concerned Scientists identifies the central focus as follows:

> ...This technology can move genes and the traits they dictate *across natural boundaries* [italics inserted by the authors] -- from one type of plant to another, from one type of animal to another, and even from a plant to an animal or an animal to a plant. *(3)*

The idea that genes moved across wide natural barriers can create risk appears to have taken hold rather broadly, based upon the concept that transcending the apparent limits of nature can lead into uncharted (and therefore dangerous) territory. As a result, proposals now exist for regulation of GMOs based upon whether the gene transfer occurs within "natural limits." A task force of the Codex Alimentarius Commission, a U.N. agency responsible for an international code for food standardization, has advanced the following language to standardize the labeling of foods derived from GMOs for purposes of international commerce:

> "Genetically modified/engineered organism" means an organism in which the genetic material has been changed through gene technology in a way that does not occur naturally by multiplication and/or any natural recombination *(4)*.

Therefore, it behooves the agricultural industries to determine whether labeling schemes for GMO foods can be based on the concept of breaching of natural reproductive barriers. To be useful, this concept must pass two tests. First, is the concept of breaching of a natural reproductive barrier definitive? Second, is there some measurable biological or chemical property that is uniquely associated with the breaching of a natural reproductive barrier through modern biotechnology, such that the breach can be routinely detected? Unfortunately, we shall see that the answer to both questions is "no."

How Definitive is the "Natural Reproductive Barrier?"

Plant breeding, a traditional activity over thousands of years, is responsible for the productive crops of today that support more than 6 billion people. Breeders take specific genetic lines that carry some desired genes, cross them with other lines carrying different genes, and select progeny with desired traits arising from both parental lines. This cycle is repeated until the needed traits are accumulated in one plant. Crosses involving sexually compatible plants are not controversial. In many cases, however, needed genes are not present in the available pool of sexually compatible germplasm. Breeders have long known how to introduce genes into a crop from some species that are related to the crop - sources that normally would not be considered sexually compatible. Tricks of the trade include using chemicals to double the number of chromosomes after crossing *(5)*; embryo rescue (growing the genetically aberrant embryo to maturity not in the mother plant, where it would spontaneously abort, but in the laboratory) *(6)*; and even somatic (vegetative) cell nuclear fusion *(7)*. Examples abound of crops that have genes from two or more "incompatible" species in their backgrounds; in fact the list comprises most of the world's major crops. Fiber quality genes in cotton *(8)*, disease resistance genes in wheat *(9)*, and stress and disease resistance genes in tomato *(10)* are but a few prominent examples. Plainly, even though these techniques do not involve the extraction, *in vitro* modification, and reinsertion of DNA into cells, they do involve breaching natural reproductive barriers. If these barriers, then, serve as the basis of a definition of GMOs, virtually all crops that feed the world today must be considered to be GMOs.

It is also true that the genome of modern plants in most cases carries segments inserted from bacteria or viruses, probably even from ancient times *(11)*. Indeed, these insertions are accepted as part of the evolution of today's flora. Viruses are highly efficient at incorporating their own DNA (usually) or RNA into the genome of the plant by asexual means, and some disease-causing bacteria can accomplish the same thing through viral plasmids. Viruses are endemic in nature, and many plants naturally contain virus-derived genes or bacterial genes *(12)*. This phenomenon also underscores the unsuitability of using "natural barriers" to define GMOs: bacterial or viral genes in higher plants demonstrate that the barrier is hardly absolute, even absent the intervention of plant breeders. Logically, then, if GMO definitions are based solely on presumed natural barriers to recombination, the GMO class could *include* most crop plants bred by traditional means, but *exclude* plants that have been genetically engineered by transfer of genes from bacteria or viruses. Neither result is the intended one.

Lastly, evidence has recently been reported for recombination of DNA, by unknown means, among fungal strains that are apparently incapable of forming a sexual stage *(13)*. Products of this poorly-defined phenomenon, which also overcomes apparent natural barriers to recombination, must also be excluded by the definition of a GMO.

Other Approaches to Defining Genetically Engineered Foods

Public concern is focused on the "unnatural" aspects of DNA modification and gene transfer across species, i.e., human manipulation of DNA with incomplete understanding of its long-term consequences. The so-called "precautionary principle" - a prohibition against introducing any new technology for which the risks are not completely understood - has been cited as a reason not to engage in such activities for fear that dangerous new plants will be created. As we have seen, the risk is inferred largely from a perception that the transfer of genes could not have occurred by natural means. By the same reasoning, little or no ecological or food safety risk is attached to naturally-occurring transfer of genes, because when natural transfer of genes is possible, it is deemed likely that Nature would have already created the recombinant organisms and tested them.

A rigorous definition of GMOs that meets societal needs, then, presents a dilemma, because the degree of risk does not correspond clearly to the type of genetic transfer, nor to the role of humans in creating the transfer. Breaching a presumed natural barrier to interspecific DNA transfer, and human intervention in recombination, are exactly the catalysts that have enabled development of today's conventional crops, which are considered safe. Furthermore, the very concept of an absolute natural barrier to DNA transfer is untenable. Genetic transfer does occur in nature between widely-separated groups of organisms, and natural events have not been considered risky. To conform with the precautionary principle, we are seemingly reduced to identifying products of concern case-by-case, relying on knowledge of the process of creation and a set of value judgments (i.e., judgments not based on analysis of scientific data) to interpret what constitutes acceptable risk.

The inherent ambiguity in defining GMOs may be reduced by beginning not with the GMO itself, but with what society wants to accomplish: reducing inferred risk without rejecting genetic progress. If one accepts the value judgments that risk might arise from (a) reliance on human judgment, which can be faulty; (b) use of a new genetic tool, which has not yet withstood a test of time; and (c) utilization of a gene that is not available through natural sexual crossing, then a serviceable definition can be fashioned. Applied to GMOs, the three putative risk factors correspond to three elements of a definition: (a) human intervention in genetic recombination [which excludes naturally-occurring recombination of any sort, including that between different classes of organisms]; (b) extraction and *in vitro* processing of DNA [which excludes all conventional breeding techniques, including those that go beyond ordinary sexual crossing between compatible plants]; and (c) sexual incompatibility of donor and receiver organisms [which excludes gene transfer that could have occurred sexually, whether or not it did so]. If a plant were to be labeled as a GMO, we propose that it must meet all three criteria. A definition might read as follows:

> A genetically engineered organism is one with recombinant DNA derived in part from DNA that was extracted from sources not sexually compatible with the target organism, modified *in vitro,* and asexually reinserted by human-directed processes.

This definition includes all crop plants that have given rise to public concern about inferred risks of modern methodology, yet it excludes genetic manipulations that have been practiced for many years and are generally accepted; and it excludes naturally-occurring recombination of all kinds. It also excludes DNA transfers between sexually compatible donor and receiver organisms (the latter could have occurred naturally and therefore pose no special risk) unless DNA from other sources is also inserted.

The definition, because of its explicit derivation from societal value judgments, is essentially a restatement of those judgments. It is useful nonetheless, because it describes in clear technical terms what products are candidates for labeling. Unfortunately, even this precise definition may fail us in the future, when advancing frontiers of science give rise to new uncertainties. A new technology, chimeraplasty, allows DNA modifications (or "repair," when unwanted mutations are present) without DNA extraction and reinsertion *(14)*. Is this modern biotechnology? Not by any of the definitions above. Will chimeraplasty be perceived to generate the same level of risk as current methods of biotechnology, and therefore require labeling based upon a revised definition? Only time will tell.

An important point is that this definition (as with all others discussed above) is based on the process by which a GMO product is made, rather than on the tangible properties of the end product. This fact strongly affects how GMOs might be detected and labeled for commerce.

Pragmatic Issues of Process-Based Regulation and Labeling

Regulation based on the process of creation of a GMO product, rather than on the product itself, presents some operational difficulties. Ideally, if the process leaves unique evidence of its use in the properties or composition of the product, then these properties can be the basis for detection. The current first-generation products of modern biotechnology largely incorporate DNA sequences that are recognizable, such as a widely used gene promoter. Indeed, PCR-based testing for such "tell-tale" DNA sequences has been approved by the European Union, but a minimum charge for such a test is U.S. $248 *(15)*.

Other detection technologies are, of course, potentially useful. If there is a unique gene product present that could have been introduced only by genetic engineering, then monoclonal antibodies, or ELISA (enzyme-linked immunosorbent assay) tests based upon monoclonals, might be used to detect it. The latter tests are an order of magnitude cheaper and more adaptable to field use than are DNA-based tests. For example, one company has developed a paper indicator strip to be immersed into an extract, and the subsequent appearance of a dark band gives a "yes-or-no" answer. This technology will not be useful, though, if the foreign gene is engineered to be inactive in the harvested tissues, as there will be no gene product to detect. Similarly, if the food is highly purified (such as seed oil or white sugar), neither specific DNA sequences nor specific proteins will be present in sufficient quantity for reliable detection, and both types of tests will fail.

Next-generation GMO products will be much more difficult to deal with than are
current products. Many different genes will be transferred into crop plants, each
with its own specific DNA sequences and gene products, for which new test kits
must be created. In some cases gene products naturally present in crop plants
will simply be increased or decreased, and ELISA tests will not distinguish these
engineered differences. (Our proposed definition excludes these plants from the
GMO category, unless there are residual foreign DNA sequences from sexually-
incompatible organisms. The primary reason for exclusion is the lack of a
natural reproductive barrier, but the operational difficulties of detecting these
types of changes provide a second, more pragmatic, reason.) More promoters
will become available for use, each with different attributes for specific
purposes; again, specific test kits will be required for each promoter sequence.
Technology is already available that allows the removal of DNA that is essential
to the recombination process but unessential thereafter, such as selectable
markers *(16)*. Tests that look for selectable markers will be useless under these
conditions.

With these types of products and technologies, there will be no one universal
means for detecting the residue of genetic engineering processes. Rather, a large
battery of tests (which might change from year to year as the tools of
biotechnology evolve) would be needed. Without prior knowledge of the nature
of the product, every test would have to be applied to every product, resulting in
exponentially escalating costs. Clearly, technology to test for evidence of
introduction of recombinant DNA will be extremely expensive, will need to be
updated frequently, and is unlikely to be sufficiently reliable. The combination
of high cost and uncertainty about the results means that current testing
procedures are badly inadequate as a basis for labeling.

The alternative to physical detection is monitoring and certification of the crop
production and handling process. Measures such as documenting the type of
seeds planted, maintaining buffer zones between fields to prevent outcrossing,
segregating harvests to prevent contamination, and preserving and tracking
identity throughout post-harvest handling will be necessary to ensure that the
conventional product remains so. Enforcement will depend, not on testing of the
endproduct, but on inspections of the farms, storage and transport facilities, and
the processing plants. This will likely also be expensive and highly intrusive,
and will require substantial restructuring of the industry to be able to comply.
For the certification to be meaningful, procedures to minimize cheating will
need to be devised and implemented.

Where Do We Go From Here?

Labeling based upon process is fundamentally different from nutritional labeling
or other labeling based on content. In the latter case, labels report tangible

properties and are value-neutral. In the former case, because they do not reflect the attributes of the product, they must reflect society's perceived need to control the process. A similar approach is embodied in the recently promulgated national certification standards for organic foods, which are based on how crops and animals are grown, rather than their intrinsic attributes. To a substantial number of consumers, the process is the most important aspect for their decision-making about the suitability of the foods they eat. (Ironically, the standards for organic foods prohibit the use of GMOs, even in the absence of a suitable definition of GMOs or a system for detecting them.) Similar process-based approaches have surfaced in recent years with demands that tuna be caught in a "dolphin-safe" manner, that cosmetics not be tested on animals, and that shoes not be manufactured by laborers working in sweatshop conditions. None of these societal values can be fulfilled by programs that rely on testing of attributes of the final product. In the case of dolphin-safe tuna, voluntary compliance and labeling has worked, and consumers do have a choice of which type of process to support with their purchases. Voluntary segregation and labeling of non-GMO foods might also work, although the cost of segregation, and of restructuring to accommodate the segregation, would necessarily be passed on to the consumers. The marketplace would soon decide the value of this consumer choice of products derived from conventional or modern biotechnology.

Literature cited

1. *What is Biotechnology?* URL
http://www.aphis.usda.gov/biotechnology/faqs.html

2. Persley, G.J.; Siedow, J.N. *Applications of Biotechnology to Crops: Benefits and Risks*; Issue Paper 12; Council on Agricultural Science and Technology: Ames, IA, 1999, p 2.

3. *What is Biotechnology?* URL http://www.ucsusa.org/biotechnology/

4. *Ad Hoc* Working Group of the *Ad Hoc* Intergovernmental Task Force on Foods Derived From Biotechnology, *General Principles for the Risk Analysis of Foods Derived from Modern Biotechnology*, July **2000**, URL
http://www.fao.org/waicent/faoinfo/economic/esn/codex/reports.htm#fbt

5. Peloquin, S.J. In *Plant Breeding II*; Frey, K.J.; Ed.: Iowa State University Press: Ames, IA, 1981; pp 117-150.

6. Simmonds, N.W.; Smartt, J. *Principles of Crop Improvement*; Blackwell Science, Ltd.: Oxford, UK, 1999; pp 262-277.

7. Cocking, E.C.; Riley, R. In *Plant Breeding II*; Frey, K. J.; Ed.: Iowa State University Press: Ames, IA, 1981; pp 85-116.

8. Meredith, Jr., W.R. In *Use of Plant Introductions in Cultivar Development, Part 1*; Shands, H.L.; Wiesner, L.E.; Eds.: CSSA Special Publication Number 17; Crop Science Society of America: Madison, WI, 1991; pp 127-146.

9. Cox, T.S. In *Use of Plant Introductions in Cultivar Development, Part 1*; Shands, H.L.; Wiesner, L.E.; Eds.: CSSA Special Publication Number 17; Crop Science Society of America, Madison, WI, 1991; pp 25-48.

10. DeVerna, J.W.; Chetelat, R.T.; Rick, C.M.; Stevens, M.A. In *Tomato Biotechnology: Proceedings of a Seminar held at the University of California, Davis. 20-22 Aug. 1986*; Nevins, D.J.; Jones, R.A. Eds.: Plant Biology, Vol. 4; Alan R. Liss, New York, pp. 27-36.

11. Bennetzen, J.L. *Curr. Opin. Plant Biol.* **1998**, Apr. 1(2), 103-108.

12. Kumar, A.; Bennetzen, J.L. *Annu. Rev. Genet.* **1999**, 33, 479-532.

13. Geiser, D.M.; Pitt, J.I.; Taylor, J.W. *Proc. Nat. Acad. Sci. USA* **1998,** 95, 388-393.

14. Zhu, T.; Peterson, D.J.; Tagliani, L.; St. Clair, G.; Baszczynski, C.L.; Bowen, B. *Proc. Nat. Acad. Sci. USA* **1999**, 96, 8768-8773.

15. Froding, J; Sigler, J. *Soya and Oilseed Bluebook Update*, **2000**, 7, URL http://www.soyatech.com/GMOtest.html, p. 3.

16. Dale, E.C.; Ow, D.W. *Proc. Nat. Acad. Sci. USA* **1991**, 88, 10558-10562.

Chapter 3

Genetically Modified Crop Approvals and Planted Acreages

V. A. Forster

Forster and Associates Consulting, 230 Steeple Chase Circle, Wilmington, DE 19808

In the past few years, genetically modified crops and their resulting food products have been in the headlines. What are these crops and the genes that have been inserted into them? Since 1992, the USDA has granted "nonregulated" status to 50 genetically modified lines of crops. Included in this number are 15 lines of corn, 5 lines of soybeans, 5 lines of cotton, 11 lines of tomato, and 4 lines of potato. The corn lines have been modified to express: tolerance to either the herbicide glyphosate (RoundUp®), or to the herbicide glufosinate-ammonium (Liberty®); resistance to the pest, european corn borer (*Ostrinia nubilalis*) ECB; or a combination of herbicide tolerance to either glyphosate or glufosinate-ammonium and ECB resistance. Four of the soybean lines have been modified to express tolerance to either glyphosate, or to (Liberty), with one type expressing modified oil (high oleic acid) content. The cotton lines have been modified to express herbicide tolerance to either glyphosate, bromoxynil or sulfonylurea, or insect resistance to the pest, pink bollworm (*Pectinophora gossypiella*) PBW and tobacco budworm (*Heliothis virescens*) TBW. One cotton line expresses bromoxynil· tolerance and PBW resistance. Nine tomato lines have been modified to delay fruit ripening. One tomato line has been modified to express resistance to the pests tomato pinworm, (*Kieferia lycopersicella*) TPW and tomato fruitworm (*Helicoverpa zea*) TFW. And one tomato line has been modified to express a lower polygalacturonase level which makes for a more meaty tomato for processing. Three modified potato lines are resistant to the Colorado Potato Beetle (*Leptinotarsa decemlineata*) CPB, and one line expresses resistance to the potato virus Y (PVY) in addition to being resistant to the CPB.

In 1998, approximately 40-50% of the total US corn planted acreage was genetically modified, with approximately one-half of the 1998 total US planted soybean acreage genetically modified. Approximately 50% of the cotton was genetically modified and approximately 5-10 % of the potato plants were genetically modified in 1998. More detail with regards to the genetic elements and approvals are given in Table 1.

Planted Acreages to Genetically Modified Crops

Figure 1 indicates the adoption rates by farmers, according to NASS and the Economic Research service, of genetically modified varieties of corn, soybean, and cotton, respectively, from 1996 to 2000. The crops are divided by trait, Bt, crops which express insect resistance that results from expression of the delta-endotoxin protein of the naturally occurring soil microbe *Bacillus thuringiensis*; and herbicide-tolerant.

Lines and Events of Crops No Longer Regulated by USDA

Table 1, Molecular Information and Approval Status of Selected Events, lists some of the major products including the crop, trait (phenotype), registrant, transformation event, inserted genes, and registration status in some of the major commodity trading partners. For a complete listing of the approximately 50 events no longer regulated by USDA, the reader is directed to the Websites listed in the References section.

Acknowledgements

The author would like to thank the following individuals for reviewing this manuscript to ensure the accuracy of information presented: Jeff Stein, Novartis Seeds; Penny Hunst, Dow AgroSciences; Dirk Klonus, Aventis Crop Sciences and Raymond Dobert, Monsanto Company.

**Percent total Crop
Acreage**

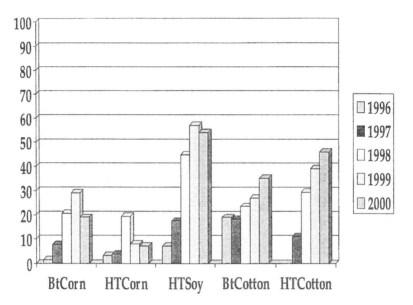

1996, 1997 and 1998 figures are from: ERS, USDA "Genetically Engineered Crops for Pest Management", 1999

1999 figures are from: NASS, USDA," Prospective Plantings, March 2000". 1999 data are a percent of harvested acres. HT crops include those derived from using both biotechnology and conventional breeding techniques

2000 figures are from: NASS, USDA, "Acreage June 2000". 2000 data are planted acres.

No **1996** data is available on HT cotton

Figure 1. Extent of Bt and herbicide-tolerant seed technologies used in corn, soybean and cotton production

Table 1: Molecular Information and Approval Status of Selected Events

Product	NatureGard®, Knockout®	YieldGard®	YieldGard®
Crop	Corn	Corn	Corn
Phenotype	ECB resistance, glufosinate tolerance	ECB resistance, glufosinate tolerance	ECB resistance
Event	Event 176	Bt11	MON 810
Petitioner	Ciba Geigy	Northrup King	Monsanto
Genotype 1			
Promoter	PEPC promoter (1 copy), Pollen promoter (1 copy)	5' CaMV35S	5' CaMV35S
Gene	Cry1A(b) from Bacillus thuringiensis ssp. kurstaki	Cry1A(b) from Bacillus thuringiensis ssp. kurstaki	Cry1A(b) from Bacillus thuringiensis ssp. kurstaki
Terminator	3' CaMV35S	nos 3' from A. tumefaciens	nos 3' from A. tumefaciens
Other	intron #9 from PEPC gene from Zea mays	IVS intron from Zea mays	hsp70 intron from Zea mays
Genotype 2			
Promoter	5' CaMV35S	5' CaMV35S	
Gene	pat from S. viridochromogenes	pat from S. viridochromogenes	
Terminator	3' CaMV35S	nos 3' from A. tumefaciens	
Other		IVS2 intron from Zea mays	
Method of transformation	microprojectile bombardment	direct DNA transfer	microprojectile bombardment
Date Nonregulated by USDA	5/17/95	1/18/96	3/15/96
Date FDA Consultation Final	1995	1996	1996
EU 90/220 Approval Status/Date	2/4/97	6/9/98 (approval for import and processing only, not for growing)	8/3/98
Japan Approval Status/Date	10/1996 (Environment); 9/1996 (Food and Feed)	10/1996 (Environment); 9/1996 (Food and Feed)	10/1996 (Environment); 5/1997 (Food); and 6/1997 (Feed)

Product	LibertyLink®	RoundUp Ready®	BollGard®, InGard®
Crop	Corn	Soybean	Cotton
Phenotype	glufosinate tolerance	glyphosate resistant	PBW resistant
Event	Events T14 and T25	GTS-40-3-2	IPC531
Petitioner	AgrEvo	Monsanto	Monsanto
Genotype 1			
Promoter	5' from CaMV35S	5' CaMV35S	5' CaMV35S
Gene	pat from S. viridochromogenes	CP4 EPSPS from Agrobacterium tumefaciens	CryIA(c) gene from Bacillus thuringensius ssp. kurstaki
Terminator	3' from CaMV35S	nos 3' from A. tumefaciens	nos 3' from A. tumefaciens or 7S 3' from soybean
Other		chloroplast transit peptide from petunia hybrida - leader	may use duplicate copy of 5" CaMV35S (enhancer)
Genotype 2			
Promoter			5' CaMV35S
Gene			nptII from E. coli Tn5
Terminator			nos 3' from A. tumefaciens or 7S 3' from soybean
Other			may use duplicate copy of 5" CaMV35S (enhancer)
Method of transformation	A. tumefaciens	particle acceleration	A. tumefaciens mediated
Date Nonregulated by USDA	6/22/95	5/19/94	6/22/95
Date FDA Consultation Final	1995	1995	1995
EU 90/220 Approval Status/Date	8/3/98 (Event T25 only)	5/0796, for import crushing and food use	pending
Japan Approval Status/Date	12/1997 (Environment); 7/1997 (Food); and 5/1997 (Feed)	3/1996 (Environment); 9/1996 (Food and Feed)	4/1997 (Environment); 5/1997 (Food); and 6/1997 (Feed)

22

References

Fernandez-Cornejo, J. and McBride, W.D.. "Genetically Engineered Crops for Pest Management in U.S. Agriculture", Economic Research Service, USDA, Agricultural Economic Report No. 786, April 2000.

"Genetically Engineered Crops for Pest Management", Economic Research Service, USDA. *Report no longer available.*

www.aphis.usda.gov/biotech/petday.html

www.ers.usda.gov

www.usda.gov/nass

http://vm.cfsan.fda.gov/~lrd/biocan.html

www.epa.gov/fedrgstr/EPA-PEST.htm

www.rki.de/GENETIC/INVEHRKEHR/INVKLIST_E.HTM

http://ss.s.affrc.go.ip/docs/sentan/eguide/commercnew.htm

Chapter 4

Insect-Resistant Transgenic Crops

J. J. Adamczyk, Jr., and D. D. Hardee

Southern Insect Management Research Unit, Agricultural Research
Service, U.S. Department of Agriculture, Stoneville, MS 38776

This chapter provides a broad overview of the development of
transgenic crops systems for plant-resistance to insects. As of
2001, the only transgenic crops that were commercially
available on a worldwide scale were those that contain the Cry
proteins from the soil bacterium, *Bacillus thuringiensis* (Bt).
The impact of Bt on non-target organisms, including
mammals and man, economic benefit for growing transgenics,
and resistance management strategies designed to preserve
this vital technology, are discussed.

Bacillus thuringiensis

Since the mid-1990's, transgenic crops have been commercially available
that control a wide range of insect species. These crops contain an insecticidal
protein from *Bacillus thuringiensis* Berliner or commonly referred to as *Bt,* a
soil bacterium that produces crystalline protein (Cry) inclusion bodies during
sporulation (*1*). These "Cry" proteins are selectively active against a multitude
of insects. By binding to receptors located in the midgut of insects, Cry proteins
form ion-selective channels in the cell membrane and causes the epithelium
cells to swell and lyse due to an influx of ions and water which leads to death in

susceptible insects (2,3). Epithelium cells of non-target organisms, such as mammals, including man, do not contain receptors for Cry proteins and thus are not affected by the toxin (4,5). Furthermore, since Cry proteins are produced as protoxins that need proteolytic activation upon ingestion, an alkaline environment (i.e. insect midgut) is crucial to allow binding to receptors. The acidic nature of mammalian digestive systems is not ideal for activation (6).

Since 1961, Bt has been used as a foliar microbial pesticide. Between 1961 and 1995, the United States Environmental Protection Agency registered 177 products that contained viable Bt (7). These formulations can contain up to 8 different Cry proteins, account for 1 – 2% of the global market (8), and are very safe to mammals as shown in Table I.

Development of Transgenic Crops

The first transgene (cry gene) was cloned and expressed in the bacterium Escherichia coli in 1981 (9). In just a few years, crops such as tomato, tobacco, and cotton plants were transformed with these cry genes (10-13). The low-toxicity profile of these transgenic crops appears to be quite similar to microbial Bt products (14). As of 2001, various crops containing Cry proteins have been registered in the United States and elsewhere and commercially available to growers to control insects as shown in Table II [see reference (15) for a thorough review of the development of transgenic Bt cotton]. To date, products that contain Cry protein(s) are the only transgenic crops designed to control lepidopteran and coleopteran pests; however, with over 100 cry genes that have been described from thousands of Bt strains identified worldwide (26), the possibilities for transforming various crops with Cry proteins seems unlimited (13). Although the first transgenic Bt plants contained only one Cry protein, experimental crops are now currently being developed that contain multiple or even hybrid cry genes that have the potential to increase the spectrum of insect control as well as decrease the chance for these pest to develop resistance to Bt (19-21; 27,28).

Impact of Cry Proteins on non-Target Organisms and Persistence in the Environment

Because current and experimental Cry protein(s) present in transgenic crops are selective against certain Lepidoptera and Coleoptera, their activity is greatly reduced against beneficial insects as shown in Table III. (30,31). As a direct consequence of reduced foliar applications being applied to transgenic crops (32), beneficial insects are not adversely affected by the broad spectrum insecticides (e.g. pyrethroids, carbamates, and organophospates) commonly used on conventional cotton and populations may actually increase (33). In addition, because Cry proteins are very selective against the target pest,

Table I. **Mammalian Toxicity Assessment of** *Bacillus thuringiensis* **as Microbial Pesticides**

Bt Strain (Microbial)	Cry Protein	Study	Findings
Kurstaki (Lepinox)	1Aa 1Ac 3Ba	Acute oral toxicity/ pathogenicity (rat)	No evidence of toxicity
Tenebrionis (San Diego)	3Aa	Acute oral toxicity (rat)	Same as above
Kurstaki (Dipel)	1Aa 1Ab 1Ac 2Aa	13-week oral (feed) (rat)	Same as above
Israelensis (h-14)	4A 4B 10A 11A 1Aa	Same as above	Same as above
Berliner	1Ab 1B	5-d human oral exposure	No adverse effects

SOURCE: Reproduced with permission from reference (*29*). Copyright 2000 Academic Press.

Table II. Transgenic Bt Crops Currently Available or Under Development: 1995 – 2001.

| Crop | Commercially Available | | Experimental | |
	Cry Protein(s)	Pest Controlled	Cry Protein(s)	Add'Pest Controlled
Cotton	1Ac[1]	Tobacco budworm, Pink bollworm, Corn earworm	1Ac + 2Ab[2] 1Ac/2Ab2[3] 1Ac/1Fa[3] 1Ac/1Ca[3]	Armyworms and Loopers, Cutworms, Cotton leaf perforator
Corn	1Ab[4] 1Ac[4] 9c[5]	European corn borer, Southwestern corn borer, Corn earworm,	1F[6] 3Bb[6] 149B1[6]	Wireworms, Corn rootworms, Fall armyworms
Potato	3A[7]	Colorado potato beetle	N/A	N/A
Soybean	N/A	N/A	1Ac[6]	Corn earworm, Loopers, Velvetbean caterpillar
Tomato	N/A	N/A	1Ac[6]	Tomato fruitworm, Loopers, Hornworms

References:
[1] (16-18), [2] (19,20), [3] (21), [4] (22), [5] (23), [6] (24), [7] (25).

beneficial insect (e.g. parasitoids) activity against secondary and occasional pests (e.g. beet armyworms) found in transgenic crops can increase or at least is maintained compared to those same pest populations found in conventional cotton as shown in Figure 1.

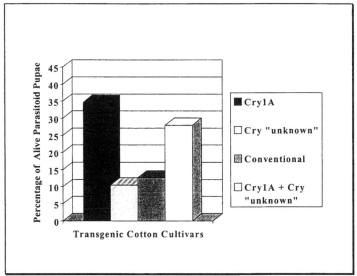

Figure 1. Parasitism of beet armyworms collected from transgenic Bt cotton by the parasitoid, Cotesia marginiventris. Cry "unknown" was from an experimental transgene (Adamczyk, unpublished).

Cry proteins appear to degrade quite rapidly in the environment. These proteins degrade in the soil at rates comparable to those used as microbial Bt insecticides (*34-37*). Thus, the low toxicity against non-target organisms (i.e. collembola, earthworms, and nematodes), combined with the rapid degradation of Cry proteins present in transgenic crops, provides an important component for controlling insect pests under ideal integrated pest management systems. However, before commercialization of a transgenic crop, studies must address the potential for flow or spread of Cry proteins to non-target plants. These studies must address the following:

- The presence of wild plant relatives (e.g. wild corn species in Mexico).
- The outcrossing potential to wild plant relatives (e.g. toxin levels in pollen, mode of reproduction).
- Potential to confer increased fitness to the recipient plant species.

Table III. Toxicity of Cry Proteins to Non-Target Organisms.

	Cry3A (Potato)	CrylAc(Cotton, Corn)	CrylAb (Corn)
Lady Bird Beetle	Practically non-toxic	Practically non-toxic: fed at 1,700x and 10,000x level in cotton pollen and nectar	Practically non-toxic NOEC>20 ppm
Collembola	NOEC>200 ppm	NOEC>200 ppm	NOEC>200 ppm
Honey Bee	Practically non-toxic to larvae	Practically non-toxic: fed at 1,700x and 10,000x level in cotton pollen and nectar	Practically non-toxic to larvae NOEC>20 ppm (larvae)
Earthworm	N/A	N/A	Practically non-toxic NOEC>20 ppm
Parasitic Wasp	Practically non-toxic	Practically non-toxic: fed at 1,700x and 10,000x level in cotton pollen and nectar	Practically non-toxic NOEC>20 ppm
Green Lace-Wing	Practically non-toxic	Practically non-toxic: fed at 1,700x and 10,000x level in cotton pollen and nectar	Practically non-toxic NOEC>16.7 ppm
Bobwhite Quail	Practically non-toxic LC_{50}>50,000 ppm (potato tubers)	Practically non-toxic	NOEC>100,000 ppm corn grain containing the CrylAb protein
Daphnia	N/A	N/A	Practically non-toxic

Table III. *Continued*

			NOEC>100 ppm of corn pollen containing Cry1Ab
Fish	N/A	N/A	No effect on channel catfish fed ground corn grain containing Cry1Ab protein (35%)

SOURCE: Reproduced with permission from reference (*29*). Copyright 2000 Academic Press.

Note: NOEC refers to the no observed effect concentration.

Economics

As usually occurs with most new technology, transgenic crops offer value to growers in a variety of ways, often depending on government commodity programs. In the case of transgenic Bt cotton, *Gossypium hirsutum* L. and Bt corn, *zea mays* L., information is limited due to the short time the technology has been available and the rapidly changing varietal structures.

Bt cotton

In 1995, the year prior to the introduction of Bt cotton, the tobacco budworm (*Heliothis virescens* F.), bollworm (*Helicoverpa zea* Boddie), and pink bollworm (*Pectinophora gossypiella* Saunders), cost growers over $250 million in lost yields (*38*). Adoption of Bt cotton in 1996 and beyond has been credited with $20-141 million gains per year for U.S. producers (*32,39,40*). The mid-south and southeastern portions of the cotton belt, i.e. areas with traditionally high populations of resistant (*41-43*) tobacco budworms, *Heliothis virescens* (F.) (Lepidoptera: Noctuidae), may benefit greatly from the use of Bt cotton (*44*), but only if the severity of these insects is high. Although (*45*) concluded that growers planting Bt cotton averaged an economic advantage of $49.80/acre per year over a 6-year period, a 4-year study in Mississippi e.g. (*46-48*) showed conclusively that the benefits of Bt is an almost absolute function of the levels of tobacco budworm infestations in a given year. These findings probably hold true for all cotton production states. Since there is no practical way to predict the severity of insect infestations before planting the new crop, growers must use previous history of infestations, concerns over secondary pest outbreaks caused by insecticide sprays, and the cost of the technology fee to aid them in making planting decisions. Although current Bt cotton varieties provide a lower level of suppression of bollworms (*49*), planting of Bt cotton provides a level of suppression often high enough to prevent the need for bollworm sprays. Even if bollworm numbers in Bt cotton dictate the need for additional control measures, bollworms can be managed with less expensive applications of insecticides than those needed for tobacco budworms (*50*).

In the arid sections of the western U.S. cotton belt, the pink bollworm is the major cotton pest, against which current varieties of Bt cotton are extremely effective (*51,52*). It is estimated that a 1-10% yield increase and a 5-65% reduction in pest control costs can be achieved using Bt cotton, but this depends on severity of pink bollworm infestations (*40*). Two factors impossible to measure from an economics standpoint are:

• The peace of mind which Bt cotton offers a producer and consultant.

- The environmental improvements resulting from decreased insecticide output due to effectiveness of Bt cotton (*49*).

Bt corn

It is estimated that the European corn borer (ECB), *Ostrinia nubilalis* (Hübner), costs farmers over $1 billion annually in yield losses and control costs (*53*). As with Bt cotton, corn economists emphasize (*54*) that the greatest value of planting Bt corn is reduced yield losses in years of significant infestations of ECB. In addition, Bt corn increases profit variability and therefore increases risk (*54*). Thus, corn farmers should probably not pay more for Bt corn than the expected value of increased yields.

Resistance Management

It is known that more than 500 species of insects and mites have developed at least some degree of resistance to insecticides (*55*), which clearly demonstrates that many arthropods have the genetic potential for rapid adaptation to chemicals in their environment. Field and laboratory studies have documented the development of resistance of several insects to spray formulations of Bt toxins. The best-known example is the diamondback moth, *Plutella xylostella* (L.), a caterpillar pest that attacks cabbage and related plants. It has shown high levels of resistance to Bt sprays in Florida, Hawaii, North Carolina, Asia, and other locations (*56*). It has also shown resistance to Bt transgenic canola plants. Researchers have already developed laboratory colonies of Colorado potato beetles, *Leptinotarsa decemlineata* (Say), European corn borers, tobacco budworms, and bollworms that are resistant to Cry-proteins. Crop protection with transgenic crops is a form of host plant resistance, such as resistance of soybean varieties to the soybean cyst nematode. Farmers are familiar with resistant crops losing their protection from pests, as happens when nematodes overcome soybean resistance, or when mildew adapts to resistant wheat varieties. While the same fate is predicted for transgenic crops, especially Bt cotton and Bt corn, the time necessary to reach economic resistance can be greatly influenced by the way growers and consultants utilize this crop, i.e. resistance management. For a detailed discussion of this topic, see reference (*49*) for cotton, and (*57*) for corn.

In order to formulate a resistance management plan, the first essential is to develop a resistance monitoring program, the aim of which is to provide timely information to be used to document resistance development, formulate alternative management options, and provide tactics to delay resistance (*58*). To date, resistance monitoring programs developed for cotton (*59-62*) and corn

(*63*) have not documented field failures of insect control in either of these crops due to development of resistance to Bt toxins. The recent identification of a gene associated with Bt resistance in a laboratory colony of tobacco budworm (*64*) was a significant step toward monitoring for field resistance to toxic proteins in Bt cotton as shown in Figure 2.

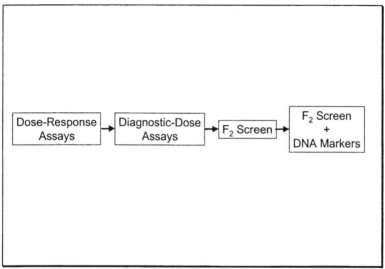

Figure 2. Evolution of resistance monitoring methods (Sumerford, unpublished).

Most scientists agree that pest insects will eventually become resistant to the Cry-proteins used in current Bt crops, especially if resistant management measures are not enforced. The tobacco budworm has a well-known reputation for developing resistance to chemical insecticides, and is currently resistant to most conventional insecticides used on cotton. However, for the time being, it is extremely susceptible to the Cry1Ac protein in Bt cotton, as is the pink bollworm. The bollworm is inherently more tolerant to this toxin, and it is likely to develop resistance faster than the tobacco budworm or pink bollworm.

The length of time that Bt cotton or corn remains effective may depend upon how well growers and pest managers follow resistance management guidelines. Improper usage dramatically decreases the effective life of a product. If Bt products are carefully used, this effectiveness may be extended for many years. But if the technology is abused, insects will quickly become resistant. Preserving the effectiveness of Bt crops is one way to keep pest management costs at the lowest level.

A logical, science-based, and proactive resistance management strategy is necessary to prevent insects from developing resistance to Bt crops in less than

10 years. All members of the corn and cotton industries should practice this strategy to slow development of resistance. Companies that sell Bt seed are required by the EPA to recommend and support insect resistance management (IRM) strategies for Bt crops. IRM is a key element of a good overall integrated pest management (IPM) program.

A resistance management concept for Bt cotton and corn, accepted by the EPA, is known as the "high dose/refuge strategy". This approach has two complementary principles:

- Bt plants must produce a high dose of Cry-toxin throughout the season.
- Effective IRM refuges must be maintained.

An IRM refuge consists of a non-Bt host crop, and it is intended to produce susceptible insects. Without a source for producing susceptible insects, the development of resistance is proportional to the dose; that is, the higher the dose, the more rapidly resistance develops. Therefore, the high dose/refuge strategy is a high-risk strategy, depending upon the availability of properly functioning IRM refuges. The high dose, or in other words high effectiveness, is very good for pest control, but it can cause the rapid development of resistance in the absence of effective IRM refuges.

Refuge regulations originally mandated for Bt corn and cotton varieties remained in effect through the 2001 growing season. When the registrations expire after the 2001 growing season, new refuge requirements will be forthcoming, but at the time of publication of this document, no final decisions had been made for the 2002 growing season and beyond. The issue will be debated before the final decision is made; recommendations could range from complete removal of Bt technology from the marketplace, to some variation of the 2001 guidelines.

References

1. Hofte, H.; Whitely, H. R. *Microbiol. Rev.* **1989**, 53, 242-255.
2. Knowles, B. H.; Ellar, D. J. *Biochem. Biophys. Acta.* **1987**, 924, 509-518.
3. English, L; Slatin, S. L. *Insect Biochem. Molec. Biol.* **1992**, 22, 1-7.
4. Hoffmann, C.; Luthy, P.; Hutter, R.; Pliska, V. *Eur. J. Biochem.* **1988**, 173, 85-91.
5. Sacchi, V. F.; Parenti, P.; Hanozet, G. M.; Giordana, B.; Luthy, P.; Wolfersberger, M. G. *FEBS Lett.* **1986**, 204, 213-218.
6. Kuiper, H. A.; Noteborn, J. M. *Food Safety Assessment of Transgenic Insect-Resistant Bt Tomatoes in Food Safety Evaluations*; OECD Workshop on Food Safety Evaluation; Oxford, England, 1996; pp. 12-15.
7. Siegel, J. P. *Journal of Invert. Pathol.* **2001**, 77, 13-21.

8. Baum, J. A.; Johnson, T. B.; Carlton, B. C. In *Methods in Biotechnology;* Hall, F. R.; Mean, J. J., Eds.; Biopesticides: Use and Delivery; Humana Press, Inc.: Totowa, NJ, 1999; Vol. 5, pp. 189-209.

9. Schnepf, H. E.; Whiteley, H. R. *Proc. Natl. Acad. Sci.* **1981**, 78, 2893-2897.

10. Fischhoff, D. A.; Bowdish, K. A.; Perlak, F. J.; Marrone, P. G.; McCormick, S. M.; Nidermeyer, J. G.; Dean, D. A.; Kusano-Kretzmer, K; Mayer, E. J.; Rochester, D. E.; Rogers, S. G.; Farley, R. T. *Bio/Technology* **1987**, 5, 807-813.

11. Vaeck, M.; Reybnaerts, A.; Hofte, J.; Jansens, S.; DeBeuckeleer, M.; Dean, C.; Zabeau, M.; Van Montagu, M.; Leemans, J. *Nature.* **1987**, 328, 33-37.

12. Perlak, F. J.; Deaton, R. W.; Armstrong, T. A.; Ruchs, R. L.; Sims, S. R.; Greenplate, J. T.; Fischhoff, D. A. *Bio/Technology.* **1990**, 8, 939-943.

13. Perlak, F. J.; Fuchs, R. L.; Dean, D. A.; McPherson, S. L.; Fischoff, D. A. *Proc. Natl. Acad. Sci.* **1991**. 88, 3324-3328.

14. Noteborn, H. P.; Rienenmann-Ploum, M. E.; van den Berg, J. H.; Alink, G. M.; Zolla, L.; Kuiper, H. A. *Proceeding of the 6th European Congress on Biotechnology*; Elsevier Science, 1994.

15. Benedict, J. H.; Altman, D. W. In *Biotechnology in Cotton;* Jenkins, J., Ed.; Oxford Press, in press.

16. MacIntosh, S. C.; Stone, T. B.; Sims, S. R.; Hunst, P. L.; Greenplate, J. T.; Marrone, P. G.; Perlak, F. J.; Fischhoff, D. A.; Fuchs, R. L. *J. Invertebr. Pathol.* **1990**, 56, 258-266.

17. Jenkins, J. N.; Parrott, W. L.; McCarty Jr., J. C. In *Proceedings of the Beltwide Cotton Conferences*; Dugger, P.; Richter, D., Eds.; National Cotton Council; Memphis, TN, 1990; 606-607.

18. Halcomb, J. L.; Benedict, J. H.; Cook, B.; Ring, D. R. *Environ. Entomol.* **1996**, 25, 250-255.

19. Adamczyk Jr., J. J; Hardee, D. D.; Adams, L. C.; Sumerford, D. V. *Journal of Econ. Entomol.* **2001**, 94, 284-290.

20. Stewart, S. D.; Adamczyk Jr., J. J.; Knighten, K. S.; Davis, F. M. *Journal of Econ. Entomol.* **2001**, 94, 752-760.

21. Sivasupramaniam, S.; Kabuye, V.; Malvar, T.; Ruschke, L.; Rahn, P.; Greenplate, J. In *Proceedings of the Beltwide Cotton Conferences*; Dugger, P.; Richter, D., Eds.; National Cotton Council; Memphis, TN, 2001; 837-840.

22. Clark, T. L.; Foster, J. E.; Kamble, S. T.; Heinrichs, E. A. *Journal of Entomol. Sci.* **2000**, 35, 118-128.

23. Jansens, S.; van Vliet, A.; C. Dickburt, C.; Buysse, L.; Piens, C.; Saey, B.; de Wulf, A.; Gossele, V.; Paez, A.; Gobel, E. *Crop Sci.* **1997**, 37, 1616-1624.

24. *EPA-Office of Pesticide Programs: Biopesticide Experimental Use Permits Approved,* URLhttp://www.epa.gov/oppbppd1/biopesticides/reg_activ/new_ eup_approv.htm.
25. Alyokhin, A. V.; Ferro, D. N. *Journal of Econ. Entomol.* **1999,** 92, 510-515.
26. Crickmore, N.; Ziegler, D. R.; Feitelson, J.; Schnepf, E.; Van Rie, J.; Lereclue, R.; Baum, J.; Dean, D. H. *Microbiol. and Molec. Biol. Rev.* **1998,** 62, 807-813.
27. Tabashnik, B. E. *Ann. Rev. Entomol.* **1994,** 39, 47-79.
28. Gould, F. *Ann. Rev. Entomol.* **1998,** 43, 701-726.
29. Betz, F. S.; Hammond, B. G.; Fuchs, R. L. *Regulatory Toxicology and Pharmacology.* **2000,** 32, 156-173.
30. Dogan, E. B.; Berry, R. E.; Reed, G. L.; Rossignol, P. A. *Journal of Econ. Entomol.* **1996,** 89, 1105-1108.
31. Sims, S. R. *Southwest. Entomol.* **1997,** 22, 395-404.
32. Gianessi, L. P; Carpenter, J. E. **1999**. *Agricultural Biotechnology: Insect Control Benefits.* National Center for Food and Agricultural Policy.
33. Riddick, E. W.; Dively, G.; Barbosa, P. *Journal of Entomol. Sci.* **2000,** 35, 349-359.
34. Palm, C. J.; Seidler, R. J.; Donegan, K. K.; Harris, D. *Plant Physiol. Suppl.* **1993,** 102, 106.
35. Palm, C. J.; Donegan, K. K.; Harris, D.; Seidler, R. J. *Mol. Ecol.* **1994,** 3, 145-151.
36. Palm, C. J.; Schaller, D. L.; Donegan, K. K.; Seidler, R. J. *Can. J. Microbiol.* **1996,** 42, 1258-1262.
37. Sims, S. R.; Holden, L. R. *Environ. Entomol.* **1996,** 25, 659-664.
38. Williams, M. R. In *Proceedings of the Beltwide Cotton Conferences*; Dugger, P.; Richter, D., Eds.; National Cotton Council; Memphis, TN, 1996; pp. 670-689.
39. Traxler, G.; Falck-Zepeda. 1999. *U.S. Ag Bio Forum,* vol. 2, pp 1-5.
40. Frisvold, G. B.; Tronstad, R.; Mortensen, J. In *Proceedings of the Beltwide Cotton Conferences*; Dugger, P.; Richter, D., Eds.; National Cotton Council; Memphis, TN, 2000; pp. 337-340.
41. Roush, R. T.; Luttrell, R. G. In *Proceedings of the Beltwide Cotton Conferences*; Dugger, P.; Richter, D., Eds.; National Cotton Council; Memphis, TN, 1987; pp. 220-224.
42. Elzen, G. W.; Leonard, B. R.; Graves, J. B.; Burris, E.; Micinski, S. *J. Econ. Entomol.* **1992,** 85, 2064-2072.
43. Hardee, D. D.; Adams, L. C.; Elzen, G. W. *Southwest. Entomol.* **2002a,** in press.
44. Stewart, S.; Reed, J.; Luttrell, R.; Harris, F. A. In *Proceedings of the Beltwide Cotton Conferences*; Dugger, P.; Richter, D., Eds.; National Cotton Council; Memphis, TN, 1998; pp. 1199-1203.

45. Oppenhuizen, M.; Mullins, J. W.; Mills, J. M. In *Proceedings of the Beltwide Cotton Conferences*; Dugger, P.; Richter, D., Eds.; National Cotton Council; Memphis, TN, 2001; pp. 862-865.

46. Cooke, F. T., Jr.; Scott, W. P.; Meeks, R. D.; Parvin, D. W. *Proc. Beltwide Cotton Prod. Res. Conf., National Cotton Council:* Memphis, TN, 2000, pp 332-334.

47. Cooke, F. T., Jr.; Scott, W. P.; Martin, S. W.; Parvin., D. W. *Proc. Beltwide Cotton Prod. Res. Conf., National Cotton Council:* Memphis, TN, 2001, pp 175-177.

48. Scott, W. P.; Cooke, F. T., Jr.; Snodgrass, G. L. *J. Cotton Sci.* **2001**, in press.

49. Hardee, D. D.; Van Duyn, J. W.; Layton, M.B.; Bagwell, R. D. Bt cotton and management of the tobacco budworm-bollworm complex; 2001, USDA-ARS-154, 37.

50. Layton, M. B. *Cotton Insect Control Guide;* Ms. Coop. Ext. Serv. Publ. 343, 36.

51. Wilson, F. D.; Flint, H. M.; Deaton, W. R.; Fischhoff, D. A.; Perlak, F. J.; Armstrong, T. A.; Fuchs, R. L.; Berberich, S. A.; Parks, N. J.; Stapp, B. R. *J. Econ. Entomol.* **1992**, 85, 1516-1521.

52. Henneberry, T. J.; Jech, L. F. *Southwest. Entomol.* **2000**, 25, 273-286.

53. Mason, C. E.; Rice, M. E.; Calvin, D. D.; Van Duyn, J. W.; Showers, W. B.; Hutchison, W. D.; Witkowski, J. F.; Higgins, R. A.; Onstad, D. W.; Dively, G. P. *European Corn Borer Ecology and Management.* Iowa State University, Ames, IA, North Central Regional Extension Publication No. 327.

54. Hurley, T. M.; Mitchell, P. D.; Rice, M. E. *What is the value of Bt corn?* Amer. Agric. Econ. Assoc.: Chicago, IL, 2001.

55. *Pest resistance to pesticides;* Georghiou, G. P.; Saito, T., Eds.; Plenum Press: NY, 1983.

56. Tabashnik, B. E.; Cushing, N. L.; Finson, N.; Johnson, M. W. *J. Econ. Entomol.* **1990**, 83, 1671-1676.

57. *Bt corn and European corn borer;* Ostlie, K. R.; Hutchison, W. D.; Hellmich, R. L. Eds.; University of Minnesota, St. Paul, MN, 1997, NCR Publication 602.

58. Andow, D. A.; Hutchison, W. D. In *Bt Corn Resistance Management;* Mellon, M; Rissler, J., Eds.; *Now or Never – Serious Plans to Save a Natural Pest Control*; Union of Concerned Scientists: Washington, DC, 1998; pp 19-66.

59. Sumerford, D. V.; Solomon, W. L.; Hardee, D. D. *Proc. Beltwide Cotton Prod. Res. Conf., National Cotton Council:* Memphis, TN, 2000, pp 1053-1055.

60. Patin, A. L.; Dennehy, T. J.; Sims, M. A.; Tabashnik, B. E.; Liu, Y-B; Antilla, L; Gouge, D.; Henneberry, T. J.; Staten, R. *Proc. Beltwide Cotton*

Prod. Res. Conf., National Cotton Council: Memphis, TN, 1999, pp 991-996.

61. Simmons, A. L.; Dennehy, T. J.; Tabashnik, B. E.; Antilla, L.; Bartlett, A.; Gouge, D.; Staten, R. *Proc. Beltwide Cotton Prod. Res. Conf., National Cotton Council:* Memphis, TN, 1998, pp 1025-1030.
62. Hardee, D. D.; Adams, L. C.; Solomon, W. L.; Sumerford, D. V. *J. Agric. Urb. Entomol.* **2002b**, in press.
63. Andow, D. A.; Olson, D. M.; Hellmich, R. L.; Alstad, D. N.; Hutchison, W. D. *J. Econ. Entomol.* **2000**, 93, 26-30.
64. Gahan, L. J.; Gould, F.; Heckel, D. G. *Science,* **2001**, 293, 857-860.

Chapter 5

Transgenic Technology for Insect Resistance: Current Achievements and Future Prospects

D. R. Walker[1], H. R. Boerma[1], J. N. All[2], and W, A. Parrott[1,*]

Departments of [1]Crop and Soil Sciences and [2]Entomology, The University of Georgia, Athens, GA 30602

Genetic engineering of plants to protect them against insects can overcome some of the obstacles that hinder conventional breeding, and can reduce costs and risks associated with insecticide use. Since the late 1980s, dozens of economically important plant species have been engineered to express heterologous genes providing insect protection. Crops expressing insecticidal crystal protein genes from *Bacillus thuringiensis* were among the first transgenic products approved for commercial use. While many of these were readily adopted by producers, concerns remain about consumer attitudes towards transgenic food, the long-term effects of planting large areas with transgenic insecticidal crops, and the ability of targeted pests to evolve resistance to expressed plant protectants.

Introduction

Pest-resistant crops can be an important component of integrated pest management (IPM), and the expression of heterologous insect resistance genes in crops can be a powerful supplement to conventional breeding that can provide effective protection against many major pests. Potential benefits include higher levels of protection, resulting in higher yields, and reduced pesticide use, resulting in lower production costs.

Plant resistance to insects has been classified as antixenosis, antibiosis, or tolerance. [1,2] *Antixenosis* (non-preference) affects insect behavior by discouraging feeding and/or oviposition. It may involve a morphological trait or the presence/absence of a phytochemical repellant or attractant. *Antibiosis* causes a detrimental effect on the life cycle or fitness of a pest. Some phytochemicals may be involved in both types of resistance. *Tolerance* refers to the ability of a host plant to withstand substantial damage from insects with no adverse effect on yield. The majority of the resistance genes which have been engineered into crops condition primarily antibiosis.

The development of insect-resistant crops using conventional approaches has been hindered by the genetic complexity of resistance and the cost and effectiveness of evaluating resistance in breeding programs. Some types of resistance to insects are qualitative, but resistance is often conditioned by a quantitative trait locus (QTL).[3] Plant breeders must often resort to agronomically inferior plant introductions as a source of resistance alleles. Resistance genes are often linked to genes affecting agronomic traits, so linkage drag becomes a problem when resistance QTLs are backcrossed into elite genotypes. Alternatively, selection for high yield often results in the inadvertent loss of the QTL alleles for resistance.[4] Minor QTLs are difficult to transfer using phenotypic selection, yet they may be essential to obtain a level of resistance equivalent to that of the resistant parent. In other cases, deliberate selection to alter chemical composition has reduced resistance to insects, as occurred in the development of low-gossypol cotton and low-glucosinolate cultivars of oilseed rape.[4] In crops such as maize and soybean, resistant hybrids and cultivars have been abandoned as higher-yielding but more susceptible hybrids and cultivars became available.[5,6] Traditional breeding approaches are especially slow in perennial crops, and are not an option in crop plants which are vegetatively propagated, or those which are sterile.

The dearth of high-yielding, insect-resistant crops has encouraged a heavy reliance on insecticides, especially in cotton and high-value horticultural crops, where damage tolerance is low.[7] Even when used responsibly, foliar insecticides can be washed off prematurely by rain, and will not control pests feeding in protected areas of the host plant. Pimentel estimated that only 0.1% of applied pesticides actually reach the target pests.[8] Furthermore, the low specificity of many insecticides makes them toxic to nontarget species, including beneficial arthropod predators and parasites of pest species, bees, and vertebrates.[7]

The potential of genetically engineering crops for insect resistance was recognized as an early commercial target, and insect-resistant cultivars of cotton and potato, and hybrids of maize engineered with a crystal (Cry) protein gene from *Bacillus thuringiensis* (Bt) were among the first transgenic crops tested in the field and approved for commercial planting in 1995.[9,10,11] It has been estimated that the use of cultivars expressing *cry* transgenes eliminated the need for insecticide applications on two million acres of maize and five million acres of cotton in the United States in 1998.[12]

Insecticidal Transgenes

The characteristics of the ideal transgene-encoded insecticide are (i) effectiveness against insects independently or as a component of an IPM system;

(ii) no adverse effects on host plant metabolism; (iii) toxicity against a wide range of pests, but harmlessness to beneficial insects; (iv) a mode of action different from those of currently used insecticides; (v) harmlessness to animals and to humans; (vi) inhibition of acquired resistance in pest populations; and (vii) compatibility with conventional breeding methods for yield and quality.[13] Due to the expense and technical difficulties involved in genetic engineering, single genes conditioning high levels of resistance are favored. The Bt Cry proteins possess most of these desirable characteristics and have received the most attention. Other types of insect resistance genes which have been tested, but not been commercially deployed, are proteinase inhibitors, lectins, α-amylase inhibitors, and chitinases.[14,15,16,17]

Cry Proteins from *Bacillus thuringiensis*

The Gram-positive bacterium, *Bacillus thuringiensis,* synthesizes parasporal crystalline inclusions during sporulation which contain crystal (Cry) proteins, or δ-endotoxins.[18] Genes for over 100 Cry proteins have been sequenced, and the proteins are classified based on amino acid homology.[19] The protoxins are 130-140 kDa (Cry1, Cry4a, and Cry4b) or 70 kDa (Cry2, Cry3, and Cry4d) in size.[20] Cry1, Cry2, Cry3, and Cry4 are structurally related and are toxic to lepidopterans, lepidopterans and dipterans, coleopterans, and dipterans, respectively. When ingested by larvae, Cry proteins are solubilized by the high pH and reducing conditions of the midgut.[21] The 130-140 kDa protoxins are proteolytically converted into 55-65 kDa toxins.[20]

Binding to the appropriate receptor on the surface of midgut epithelial cells induces a conformational change in the toxin, and this is followed by irreversible insertion of part of the toxin into the membrane of a columnar cell to form an ion channel or pore.[20, 22] Pore formation may involve a tetrameric association of Cry proteins, with the participation of a single domain I α-helix from each of the toxin molecules.[23] These pores permit a redistribution of cations between the midgut lumen and cell cytoplasm, disrupting the transmembrane K^+ gradient needed to drive amino acid transport into the columnar epithelial cells, ultimately resulting in osmotic cell lysis.[24]

The toxicity of each Cry protein type is limited to one or two insect orders, and is nontoxic to vertebrates and many beneficial arthropods. This high specificity combined with high efficacy against target pests has made Bt-based bioinsecticides popular. They account for approximately 90% of all bioinsecticides, but only 2% of all insecticides used.[25] Widespread use has been limited by the cost of of Bt preparations relative to that of many other insecticides, and by the short persistence of many formulations in the field.[26,27,28] Commercial Bt products, like all applied insecticides, are of limited value for the control of endophytic pests such as borers, which feed inside host plant tissues. In contrast, expression of Cry proteins in vulnerable plant tissues ensures toxin ingestion by insects feeding in protected tissues and by early instar larvae, which are more sensitive than older insects.

Cry proteins are lethal to sensitive pests at even a few parts per million in the diet, making them 300 times more potent than synthetic pyrethroids.[18,29] No other resistance transgene candidates condition such a high level of resistance at low levels of expression (e.g., even at only 0.001% of total soluble protein, some resistance is observed). Tobacco and tomato were the first plants transformed with a cry gene.[30,31,32] Since then, dozens of crop species have been engineered with one or more cry genes.[33,34,35] The most commonly used genes have been cry1Aa, cry1Ab, cry1Ac, cry3A, and Cry9.

Cry protein expression in plants

Early transformation work made it clear that native cry genes were poorly expressed in plants, due largely to the high percentage of A + T residues, improper codon usage, sequences in the coding region that acted as polyadenylation signals in plants, and/or reduced transcript stability.[36] Portions of the coding sequences were modified to improve transcription and transcript stability in plants by eliminating A + T stretches in the native sequence and by matching the codon usage pattern of plant genes, increasing the expression of Cry1Ab and Cry1Ac proteins in cotton 100-fold.[37,38] Expression levels of a modified cry3A gene were correlated with survival and growth of Colorado potato beetle larvae on transgenic plants.[39,40] Transformants on which complete mortality of neonate larvae occurred expressed the endotoxin as 0.002% to 0.3% of total leaf protein. Expression at levels in excess of 0.005% also reduced defoliation by adult beetles and oviposition by adult females.[40]

An alternative approach to the gene expression incompatibilities between bacteria and plants has been to transform the chloroplast genome.[41] Transgenic tobacco chloroplasts expressed an unmodified cry1Ac coding sequence as 3-5% of the soluble protein in the foliage. This approach has not been widely adopted due to the difficulty of transforming chloroplasts in most crops.

Proteinase Inhibitors

A number of proteinase (or protease) inhibitors (PIs) adversely affect insects by interfering with protein digestion. These PIs form stoichiometric complexes with specific proteolytic enzymes.[42] Serine PIs, which are active against a variety of herbivores, and cysteine PIs, which have an effect on several species of coleopteran and hemipteran insects, have received the most attention.[43] Serine PIs act as competitive inhibitors and stimulate hyperproduction of proteolytic enzymes, thus diverting essential amino acids away from synthesis of other proteins and adversely affecting growth and development.[44,45]

Investigations of PIs began at about the same time as the work with Cry proteins, and tobacco plants expressing the cowpea trypsin inhibitor (CpTI) were reported in 1987.[46] Two advantages of many PIs over Cry proteins are that (i) native coding sequences do not need to be re-designed to obtain high levels of expression in plants, and (ii) some, like CpTI, are effective against pests in several orders of insects. [4,47] PIs with wound-inducible promoters may be good candidates for transgene-mediated insect resistance.[48] Proteinase inhibitor and chitinase genes of insect origin have also been engineered into plants.[35] Insects tend to be most sensitive to PIs originating either from their own species or from non-host plants.[49]

In the initial work with CpTI, the PI was expressed in tobacco leaves as up to 1% of total soluble protein, providing protection against early instars of the tobacco budworm (*Heliothis virescens*).[14] The trypsin/chymotrypsin inhibitor II genes of tomato and potato, and tomato inhibitor I (primarily a chymotrypsin inhibitor) were also engineered into tobacco.[50] These initial studies were encouraging and suggested that PIs could be a useful alternative to Cry proteins. Subsequent studies, however, have revealed that several pest species are able to adapt to trypsin inhibitors in their diet by secreting inhibitor-resistant trypsin(s).[51,52,53] Field tests of tobacco expressing CpTI showed that resistance was variable from one trial to another, and was influenced by environmental conditions, plant age, and heterogeneity of insect populations.[54] Variable results also have been reported for other plants expressing heterologous PIs.[55]

To date, efforts to express heterologous protease inhibitors in plants have lagged behind work with Bt transgenics. Variation in the sensitivity of different insect pests to PIs has hindered efforts to obtain plants with a high level of resistance to multiple pests.[54] In addition, PIs are less toxic and slower-acting than Cry proteins and many synthetic insecticides. Furthermore, sub-optimal levels of expression can actually increase insect damage in some cases.[56] Combining multiple PIs or chemical modification to improve binding ability might improve the efficacy of PIs. Some combinations of serine and cysteine PIs exhibit synergistic toxicity to certain pests.[57]

Lectins and Other Compounds

The snowdrop lectin increased aphid resistance when expressed in both tobacco and potato.[54,58] Lectins are carbohydrate-binding proteins abundant in the seeds and storage tissues of some plant species, and occurring at lower levels in many other plant tissues. Many are toxic to insects, including sucking insects which cannot be controlled by PIs or Cry proteins, but some of these are also toxic to vertebrates.[59] The mechanism of action appears to involve specific binding to glycoconjugates in the insect midgut.[33] Resistance to sucking insects can be improved by coupling a lectin coding sequence to a sap-specific promoter.[54] Lectin coding sequences usually do not require modifications to obtain useful levels of expression in different plant species.

Several other classes of chemicals, such as chitinases, α-amylase inhibitors, and alkaloids are also of interest due to their toxic or antinutritive effects on insects. Chitinases of both plant and animal origin have been introduced into plants.[33,60] Since chitinases are involved in both molting and digestion in arthropods, they can disrupt the normal growth and development of insect pests exposed to them at abnormal times or in excessive quantities. Enhanced resistance to an aphid, a beetle, and a caterpillar have been reported in transgenic plants.[33,61] The demonstration that a heterologous chitinase gene can enhance the effectiveness of a Bt toxin suggests that the effects of co-expression of chitinase and cry transgenes should be investigated further.[61]

The α-amylase inhibitor (α-AI) of the common bean provides antibiosis resistance to many coleopteran pests in the family Bruchidae which are serious pests of stored seeds in developing countries. Expression of the bean α-AI gene in pea or Azuki bean conferred resistance to some, but not all species of bruchid beetles.[62] A cholesterol oxidase gene from *Streptomyces* conferred resistance to boll weevil (*Agrotis ipsilon*) larvae when expressed in tobacco, and has an LD_{50} to boll weevil larvae that is similar to that of sensitive lepidopteran species to Cry proteins.[63]

During vegetative growth, some strains of Bt produce vegetative insecticidal proteins (VIPs), which are unrelated to the Cry proteins.[64] The Vip3A protein is toxic to a wide range of lepidopteran pests, including species like armyworms (*Spodoptera* spp.) which are relatively resistant to Cry proteins.[65] VIPs have not yet been exploited for genetic engineering of plants, but may prove a useful alternative or supplementary source of resistance in the future.

Regulatory Approval

Approval of transgenic crops for commercialization in the United States is regulated by three federal agencies: the U. S. Department of Agriculture (USDA), the Environmental Protection Agency (EPA), and the Food and Drug Administration (FDA).[66] The USDA Plant and Animal Health Inspection Service oversees field testing of biotech seeds and plants, and evaluates potential risks posed to the environment. The pesticidal properties of engineered plants are determined by the EPA, and the safety of all plant material intended for food or animal feed is evaluated by the FDA. Regulatory oversight is constant throughout the development and testing of genetically engineered plants, and there are 10 separate points at which federal regulators can either delay or halt development. All decisions are based on data, and there are six specific opportunities for the public to obtain the data and register any concerns about the material under evaluation.

Potential Risks of Using Transgenic Insecticidal Crops

Risks posed by widespread use of transgenic crops vary according to the biology of the crop itself, the biology of the pests, and whether the foreign protein is toxic or allergenic to mammals. Among the concerns which have received attention are the potentials for transgene escape into weed populations, for insensitive or secondary pests to proliferate as a result of reduced applications of broad-spectrum insecticides, for the accumulation of Cry proteins in the soil, for unanticipated allergic reactions in humans, and for insect populations to become resistant to Cry proteins.

The likelihood that a transgene which increases fitness would be accidentally transferred to sexually compatible weeds depends on the degree to which outcrossing occurs in the crop species, and whether there are weedy relatives in the vicinity. *Brassica* species would pose a greater risk, for example, than either soybean, which is almost entirely self-pollinated, or maize, which is grown near a wild relative only in Mexico. Fields of Bt crops are closely monitored for signs of resistant insect populations, and studies are underway in Bt cotton to determine the impact of reduced insecticide use on populations of Bt-insensitive pests such as stink bugs.[67] The large quantities of Cry proteins produced by transgenic crops, especially when a constitutive promoter has been used, has raised concerns about the persistence and effects of Cry protein residues in the soil.[68] Cry protein concentrations appear to decline rapidly for about 14 days, after which they drop at a slower rate.[69]

Resistance Management

The widespread adoption of crops expressing *cry* genes has raised concerns that Bt-resistant pest populations will evolve. Although this is a potential problem with any type of insecticide, selection pressure would be higher with Bt crops in which the protein is expressed constitutively and continuously, especially if a large proportion of the crops in a region is planted with crops expressing the same or related transgenes. Insect pests which have developed resistance to one or more Cry proteins in the laboratory include *Plodia interpunctella*, *Plutella xylostella*, *Heliothis virescens*, *Leptinotarsa decemlineata*, *Ostrinia nubilalis*, *Spodoptera exigua*, *S. littoralis*, and *Trichoplusia ni*.[70] Field populations of *P. xylostella* resistant to topical applications of Bt have occurred in several locations around the world.[71,72] Most resistant strains exhibit some degree of cross-resistance to structurally related Cry toxins.[22] Resistance in most of these strains involves reduced binding to the receptor(s) and is functionally recessive, so heterozygotes are nearly as sensitive as susceptible homozygotes.[22,73]

Wearing and Hokkanen reviewed the factors which should be assessed in the strategic deployment of crops genetic engineered with a Bt transgene.[74] Important considerations include biological and cultural aspects of the crop (annuals vs. perennials, production practices, size of individual fields, availability of alternative host plants, and potential for intercrossing with wild relatives), and pest complex (types of pests, sensitivity of various pests to the expressed toxin, polyphagy vs.

oligophagy, mobility, pest density in the crop, mating behavior, availability of natural and cultivated refugia, and abundance of natural enemies). Optimal deployment strategies may vary from one region to another.

Acccordingly, numerous resistance management strategies have been proposed, and most can be classified as either cultural or genetic. Functionally recessive resistance is the basis for the currently implemented "high dose - refugium" approach, which attempts to kill at least 95% of the heterozygotes through high levels of expression, while promoting mating opportunities between resistant individuals and susceptible insects from nearby refugia. Refugia are areas within fields seeded to plants that do not express a Bt gene and which are not treated with any Bt preparations. Modeling studies show that mortality of heterozygotes, which are the most common carriers of a resistance allele, has the greatest influence on the evolution of resistance.[75] The complex interactions between each crop species and its suite of pests, and among the agroecosystems characteristic of a particular region must be taken into consideration in order to develop effective management strategies.[74] For example, in areas where Bt cotton is planted, there are restrictions on the amount of Bt maize that can be planted.[76]

A potential problem with the "high dose" concept is that most crops have multiple pests, and an expression level that is acutely toxic to the primary pest(s), may not control other less sensitive pests. Furthermore, Cry protein levels can exhibit temporal and spatial variation, as well as environmental influence.[77] Mating opportunities between resistant individuals from transgenic fields and susceptible individuals from adjacent refuges may be limited by asynchronous development if the effects of the Cry toxin are chronic rather than acute.[78]

The use of tissue- or time-specific promoters which would reduce exposure of insects to toxins might also help to delay the evolution of resistant pest populations.[79] The widely used CaMV 35S constitutive promoter puts uninterrupted selection pressure on sensitive insect species. A promoter which drives more selective expression may provide the resistance needed to protect yields, while reducing unnecessary exposure of insects to the toxin. Tissue-specific promoters might increase the survival of acceptable numbers of Bt-susceptible insects on transgenic plants, and some could be used to reduce or eliminate expression in tissues harvested for food.

The level and breadth of pest control could potentially be increased by pyramiding or stacking resistance genes and/or transgenes in crops to be deployed. Some combinations of Cry proteins would be of limited value due to cross-resistance, and stacking ideal combinations of *cry* transgenes may be restricted by intellectual property rights.[80] Co-expression of compatible Cry proteins can be quite effective, however. Stewart et al. demonstrated that cotton lines expressing both Cry1Ac and Cry2Ab were significantly more effective against semi-tolerant pests than another cotton line expressing only Cry1Ac. This was true for both lethal and sublethal effects. Alternatively, a *cry* transgene could be co-expressed with another transgene encoding a toxin with a completely different mode of action. Studies investigating this tactic have yielded mixed results, reflecting the complex interactions involved. MacIntosh et al. found that several proteinase inhibitors enhanced the toxicity of three different Cry proteins to sensitive pests, whereas Tabashnik et al. found no synergism between two PIs and Bt towards diamondback moth.[82,83] Zhang et al. reported that the soybean trypsin inhibitor, in combination with low levels of a Cry protein, reduced larval growth in *Helicoverpa armigera*, but higher levels of the Cry protein failed to increase larval mortality.[84] Santos et al. found that *Arabidopsis thaliana* plants expressing Cry1Ac and CpTI were actually less resistant than similar plants expressing only Cry1Ac.[85] PI-induced changes in the midgut protease composition may have accelerated degradation of the Cry protein.[84]

Meade and Hare found an additive effect of applied Bt insecticides and host plant resistance in celery, suggesting that pyramiding the two types of resistance genes could enhance resistance.[86] This strategy was investigated in soybean by Walker et al., who combined a *cry1Ac* transgene with resistance alleles at two QTLs from the plant introduction, PI 229358.[87] Detached leaves from transgenic plants carrying a resistance allele at a major QTL on linkage group M were found to be more resistant to soybean looper than leaves from transgenic plants lacking the QTL allele. Although this additive effect was not seen with corn earworm, it may have been due the greater sensitivity of this pest to the Cry1Ac protein.

Commercialization of Bt Hybrids and Cultivars

Among the first transgenic crops approved for commercial production in the USA were potato expressing a Cry3A toxin (Newleaf® from Monsanto), maize expressing Cry1Ab (Maximizer® from Syngenta), and cotton expressing Cry1Ac (Bollgard® from Monsanto). The development of transgenic crops has been hindered by the expense and technical difficulties associated with transformation, and the cost of demonstrating that a transgenic cultivar or hybrid provides acceptable levels of resistance in multiple environments and meets safety standards.

The adoption of transgenic insect-protected crops by producers has been influenced by performance, levels of insect infestations, consumer attitudes about transgenic food and fiber crops, and market economics. Widespread planting of Bt crops has thus far been limited to maize and cotton.[5] Bt maize acreage in the USA peaked in 1999 at 26%, then dropped to 19% in 2000, following two years of unusually light European corn borer infestations. The primary benefit of the use of Bt maize hybrids has been increased yields, since producers rarely try to control European corn borer with insecticide applications.

Acreage planted in Bt cotton has increased every year since 1996, and 39% of the 15 million acres planted in the USA in 2000 carried a Bt gene.[5] In four states, more than 70% of the cotton crop was planted with Bt cultivars, while plantings in the major production states of Texas and California have been limited by a lack of appropriate cultivars for the former state, and by a combination of restrictive laws and relative unimportance of lepidopteran pests in the latter.

Monsanto's "stacked" cultivars of cotton carrying both *cry1Ab* and a gene conditioning glyphosate tolerance have proven popular with growers. The popularity of Bt cotton in many states is largely due to the dual benefits of increased yield and the reduced need for insecticide applications, and has occurred in spite of occasional grower dissatisfaction in areas where unusually high pressure from pests has occurred.

Adoption of Bt potato cultivars has been very limited, and only 3% of the potatoes grown in 2000 expressed a Bt transgene.[5] Factors accounting for this include the need to use applied insecticides to control pests that are not sensitive to the expressed Cry protein, consumer concerns about transgenic foods, and the concurrent introduction of a highly effective and affordable insecticide. In addition, the primary pest targeted is the Colorado potato beetle, which is not a major pest in the northwestern U.S. production areas. Finally, aphids, which are vectors of two major viruses, are not sensitive to Cry3a, so applications of broad-spectrum insecticides are frequently necessary to control them on Bt cultivars. More recent transgenic potato cultivars have the *cry3a* stacked with genes for virus resistance, thus reducing the need for aphid control.

Future Prospects

Strategies for protection of crop plants against insects will continue to evolve as additional sources of resistance are identified and breeding strategies become more sophisticated due to further advances in transgenic and marker-assisted selection technology. Due to the dynamic evolutionary relationship between insects and their hosts, it remains likely that economic and sustainable levels of resistance to insects will require that resistant crops be deployed within the context of appropriate agricultural practices.

References

1. Painter, R. H. *Insect Resistance in Crop Plants*; Macmillan: New York, 1951.
2. Kogan, M.; Ortman, E. E.; *Bull. Entomol. Soc. Am.* **1978**, *24*, 175-176.
3. Duck, N.; Evola, S. In *Advances in Insect Control*; Carozzi, N., Koziel, M., Eds; Taylor and Francis: London, 1997; p 1.
4. Gatehouse, A. M. R.; Hilder, V. A.; Boulter, D. *Plant Genetic Manipulation for Crop Protection*; CAB International: Wallingford, 1992.
5. Carpenter, J. E.; Gianessi, L.P. *Agricultural Biotechnology: Updated Benefit Estimates*; National Center for Food and Agricultural Policy: Washington, DC, 2001.

48

6. Boethel, D. J. In *Global Plant Genetic Resources for Insect Resistant Crops*; Clement, S. L., Quisenbury, S. S., Eds.; CRC Press: Boca Raton, 1999; p 101.
7. Pimentel, D.; Greiner, A. In *Techniques for Reducing Pesticide Use*; Pimentel, D., Ed.; John Wiley: Chichester, U.K., 1997; p 51.
8. Pimentel, D. *J. Agric. Environ. Ethics* **1995**, *8*, 17-29.
9. Perlak, F. J.; Deaton, R. W.; Armstrong, T.A.; Fuchs, R. L.; Sims, S. R.; Greenplate, J. T.; Fischhoff, D. A. *Bio/Technology* **1990**, *8*, 939-943.
10. Perlak, F. J.; Stone, T. B.; Muskopf, Y. M.; Petersen, L. J.; Parker, G. B.; McPherson, S. A.; Wyman, J.; Love, S.; Reed, G.; Biever, D.; Fischhoff, D. A. *Plant Mol. Biol.* **1993**, *22*, 313-321.
11. Koziel, M. G.; Beland, G. L.; Bowman, C.; Carozzi, N.B.; Crenshaw, R.; Crossland, L.; Dawson, J.; Desai, N.; Hill, M.; Kadwell, S.; Launis, K.; Lewis, K.; Maddox, D.; McPherson, K.; Meghji, M. R.; Merlin, E.; Rhodes, R.; Warren, G. W.; Wright, M.; Evola, S. V. *Bio/Technology* **1993**, *11*, 194- .
12. Brower, V.; Dorey, E.; Fox, J.; Hodgson, J.; Saegusa, A.; Spillma, I. *Nature Biotechnol.* **1999**, *17*, 735.
13. Boulter, D. *Phytochem.* **1993**, *34*, 1453.
14. Hilder, V. A.; Gatehouse, A. M. R.; Sheerman, S. E.; Barker, R. F.; Boulter, D. *Nature* **1987**, *333*, 160.
15. Hilder, V. A.; Powell, K. S.; Gatehouse, A. M. R.; Gatehouse, J. A.; Gatehouse, L. N.; Shi, Y.; Hamilton, W. D. O.; Merryweather, A.; Newell, C. A.; Timans, J. C. et al. *Transgen. Res.* **1995**, *4*, 18.
16. Shade, R. E.; Schroeder, H. E.; Pueyo, J. J.; Tabe, L. M.; Murdock, L. L.; Higgins, T. J. V.; Chrispeels, M. J. *Bio/Technology* **1994**, *12*, 793.
17. Gatehouse, A. M. R.; Davidson, G. M.; Newell, C. A. *Mol. Breed.* **1997**, *3*, 49-63.
18. Feitelson, J. S.; Payne, J.; Kim, L. *Bio/Technology* **1992**, *10*, 271-275.
19. Crickmore, N.; Zeigler, D. R.; Feitelson, J.; Schnepf, E.; Van Rie, J.; Lereclus, D.; Baum, J.; Dean, D. H. *Microbiol. Mol. Biol. Rev.* **1998**, *62*, 807-813.
20. Knowles, B. H.; Dow, J. A. T. *Bioessays* **1993**, *15*, 469-476.
21. Aronson, A. I.; Shai, Y. *FEMS Microbiol. Letters* **2001**, *195*, 1-8.
22. Schnepf, E.; Crickmore, N.; Van Rie, J.; Lereclus, D.; Baum, J.; Feitelson, J.; Zeigler, D. R.; Dean, D. H. Microbiol. *Mol. Biol. Rev.* **1998**, *62*, 775-806.
23. Vié, V.; Van Mau, N.; Pomarède, P.; Dance, C.; Schwartz, J. L.; Laprade, R.; Frutos, R.; Rang, C.; Masson, L.; Heitz, F.; Le Grimellec, C. *J. Membrane Biol.* **2001**, *180*, 195-203.
24. Karim, S.; Gould, F.; Dean, D. H. *Current Microbiol.* **2000**, *41*, 214-219.
25. Lambert, B.; Peferoen, M. *BioScience* **1992**, *42*, 112.
26. Adang, M. J. In *Biotechnology for Biological Control of Pests and Vectors*; Maramosch, K., Ed.; CRC Press: Boca Raton, 1991; p 3.
27. Cannon, R. J. C. *Pesticide Sci.* **1993**, *37*, 331.
28. Marrone, P. G. In *Biotechnology and Integrated Pest Management*; Persley, G. J., Ed; CAB International: Wallingford, 1996; p 150.

29. MacIntosh, S. C.; Kishore, G.; Perlak, F. J.; Marrone, P. G.; Stone, T. B.; Sims, S. R.; Fuchs, R. L. *J. Agric. Food Chem.* **1990**, *38*, 1145-1152.
30. Barton, K. A.; Whiteley, H. R.; Yang, N.-S. *Plant Physiol.* **1987**, *85*, 1103-1109.
31. Vaeck, M. A.; Reynaerts, A.; Höfte, H.; Jansens, S.; de Beukleer, M; Dean, C.; Zabeau, M.; van Montagu, M.; Leemans, J. *Nature* **1987**, *328*, 33-37.
32. Fischhoff, D. A.; Bowdish, K. S.; Perlak, F. J.; Marrone, P. D.; McCormick, S. M.; Niedermeyer, J. G.; Dean, D. A.; Kusano-Kretzmer, K.; Mayer, E. J.; Rochester, D. A.; Rogers, S. G.; Fraley, R. T. *Bio/Technol.* **1987**, *5*, 807-813.
33. Jouanin, L.; Bonnade-Bottino, M.; Girard, C.; Morrot, G.; Gibaud, M. *Plant Sci.* **1998**, *131*, 1-11.
34. Mazier, M.; Pannetier, C.; Tourneur, J.; Jouanin, L.; Gibaud, M. *Biotechnol. Ann. Rev.* **1997**, *3*, 313-347.
35. Schuler, T. H.; Poppy, G. M.; Kerry, B. R.; Denholm, I. *Trends Biotechnol.* **1998**, *17*, 210.
36. Murray, E. E.; Rocheleau, M.; Eberle, M.; Stock, C.; Sekar, V.; Adang, M. J. *Plant Mol. Biol.* **1991**, *16*, 1035-1050.
37. Adang, M. J. In *Biotechnology for Biological Control of Pests and Vectors*; Maramorosch, K. Ed.; CRC Press: Boca Raton, **1991**; p. 3.
38. Perlak, F. J.; Fuchs, R. L.; Dean, D. A.; McPherson, S. L.; Fischhoff, D. A. *Proc. Natl. Acad. Sci. USA* **1991**, *88*, 3324-3328.
39. Adang, M. J.; Brody, M. S.; Cardineau, G.; Eagan, N.; Roush, R. T.; Shewmaker, C. K.; Jones, A.; Oakes, J. V.; McBride, K. E. *Plant Mol. Biol.* **1993**, *21*, 1131-1145.
40. Perlak, F. J.; Stone, T. B.; Muskopf, Y. M.; Petersen, L. J.; Parker, G. B.; McPherson, S. A.; Wyman, J.; Love, S.; Reed, G.; Biever, D.; Fischhoff, D. A. *Plant Mol. Biol.* **1993**, *22*, 313-321.
41. McBride, K. E.; Svab, Z.; Schaaf, D. J.; Hogan, P. S.; Stalker, D. M.; Maliga, P. *Bio/Technology* **1995**, *13*, 362-365.
42. Marchetti, S.; Delledonne, M.; Fogher, C.; Chiabà, C.; Chiesa, F.; Savazzini, F.; Giordano, A. *Theor. Appl. Genet.* **2000**, *101*, 519-526.
43. Koiwa, H.; Bressan, R. A.; Hasegawa, P. M. *Trends Plant Sci.* **1997**, *2*, 379-384.
44. Broadway, R. M. *J. Insect Physiol.* **1995**, *41*, 107-116.
45. Broadway, R. M.; Duffey, S. S. *Insect Physiol.* **1986**, *32*, 827-833.
46. Hilder, V. A.; Gatehouse, A. M. R.; Sheerman, S. E.; Barker, R. F.; Boulter, D. *Nature* **1987**, *330*, 160-163.
47. McManus, M. T.; Burgess, E. P. J. *J. Insect Physiol.* **1995**, *41*,731-738.
48. Duan, X.; Li, X.; Xue, Q.; Abo-El-Saad, M.; Xu, D.; Wu, R. *Nature Biotechnol.* **1996**, *14*, 494-498.
49. Jongsma, M. A.; Stiekema, W. J.; Bosch, D. *Trends Biotechnol.* **1996**, *14*, 331-333.
50. Johnson, R.; Narvaez, J.; An, G.; Ryan, C. *Proc. Natl. Acad. Sci. USA* **1989**, *86*, 9871-9875.

50

51. Bolter, C.; Jongsma, M. *J. Insect Physiol.* **1995**, *41*, 1071-1078.
52. Broadway, R. M. *Can. J. Plant Pathol.* **1996**, *32*, 39-53.
53. Wu, Y.; Llewellyn, D.; Matthews, A.; Dennis, E. S. *Mol. Breed.* **1997**, *3*, 371-380.
54. Gatehouse, A. M. R.; Powell, K. S.; Brough, C.; Hilder, V. A.; Hamilton, W. D. O.; Newell, C. A.; Merryweather, A.; Boulter, D.; Gatehouse, J. A. In *The Production and Uses of Genetically Transformed Plants*; Harrison, B. D., Leaver, C. J., Eds.; The Royal Soc./Chapman & Hall: London, 1994, p 91-98.
55. Johnson, R.; Narvaez, J.; An, G.; Ryan, C. *Proc. Natl. Acad. Sci. USA* **1989**, *86*, 9871-9875.
56. De Leo, F.; Bonadé,-Bottino, M. A.; Ceci, L. R.; Gallerani, R.; Jouanin, L. *Plant Physiol.* **1998**, *118*, 997-1004.
57. Oppert, B.; Morgan, T.; Cubertson, C.; Kramer, K. *Comp. Biochem. Physiol.* **1993**, *105C*, 379-385.
58. Down, R. E.; Gatehouse, A. M. R.; Hamilton, W. D. O.; Gatehouse, J. A. *J. Insect Physiol.* **1996**, 42, 1035-1045.
59. Clarke, E. J.; Wiseman, J. *J. Agricult. Sci., Cambridge* **2000**, *134*, 125-136.
60. Gatehouse, A. M. R.; Davison, G. M.; Newell, C. A.; Merryweather, A.; Hamilton, W. D. O.; Burgess, E. P. J.; Gilbert, R. J. C.; Gatehouse, J. A. *Mol. Breed.* **1997**, 3, 49-63.
61. Ding, X.; Gopalakrishnan, B.; Johnson, L. B.; White, F. F.; Wang, X.; Morgan, T. D.; Kramer, K. J.; Muthukrishnan, S. *Transgenic Res.* **1998**, *7*, 77-84.
62. Shade, R. E.; Schroeder, H. E.; Pueyo, J. J.; Tabe, L. M.; Murdock, L. L.; Higgins, T. J. V.; Chrispeels, M. J. *Bio/Technol.* **1994**, 12, 793-796.
63. Purcell, J. P.; Greenplate, J. T.; Jennings, M. G. *Biochem. Biophys. Res. Commun.* **1993**, *196*, 1406-1413.
64. Estruch, J. J.; Warren, G. W.; Mullins, M. A.; Nye, G. J.; Craing, J. A.; Koziel, M. G. *Proc. Natl. Acad. Sci. USA* **1996**, 5389-5394.
65. Estruch, J. J.; Carozzi, N. B.; Desai, N.; Duck, N. B.; Warren, G. B.; Koziel, M. G. Nature Biotechnology **1997,** 15, 137-141.
66. Coordinated Framework for Regulation of Biotechnology, 51 Fed. Reg. 23,302, 23,302, 1986.
67. Bundy, C. S.; McPherson, R. M. *J. Econ. Entomol.* **2000**, *93*, 697-706.
68. Palm, C. J.; Donnegan, K.; Harris, D.; Seidler, R. *J. Mol. Ecol.* **1994**, *3*, 145-151.
69. Palm, C. J.; Schaller, D. L.; Donnegan, K. K.; Seidler, R. J. *Can. J. Microbiol.* **1996**, *42*, 1258-1262.
70. Frutos, R.; Rang, C.; Royer, M. *Crit. Rev. Biotechnol.* **1999**, *19*, 227-276.
71. Kirsch, K.; Schmutterer, J. *J. Appl. Entomol.* **1988**, *105*, 249-255.
72. Tabashnik, B. E.; Cushing, N. L.; Finson, N.; Johnson, M. W. *J. Econ. Entomol.* **1990**, *83*, 1671-1676.
73. Tabashnik, B. E. *Annu. Rev. Entomol.* **1994**, *39*, 47-79.

74. Wearing, C. H.; Hokkanen, H. M. T. In *Biological Control: Benefits and Risks*; Hokkanen, H. M. T., Lynch, J. M., Eds; Cambridge Univ. Press: New York, 1995; p 236.
75. Roush, R. T. In *Advances in Insect Control: The role of transgenic plants*; Carozzi, N., Koziel, M., Eds; Taylor and Francis, Ltd.: London, 1997; p 271-294.
76. Van Duyn, *plymouth.ces.state.nc.us/pubs/btcorn99.html*. **1999**.
77. Greenplate, J. T. *J. Econ. Entomol.* **1999**, *92*, 1377-1383.
78. Gould, F. *Biocontrol Sci. Technol.* **1994**, *4*, 451-461.
79. Gould, F. *Ann. Rev. Entomol.* **1998**, *43*, 701-726.
80. McGaughey, W. H. Biocontrol Sci. Technol. **1994**, *4*, 427-435.
81. Stewart, S. D.; Adamczyk, J. J.; Knighten, K. S.; Davis, F. M. *J. Econ. Entomol.* **2001**, *94*, 752-760.
82. MacIntosh, S. C.; Stone, T. B.; Jokerst, R. S.; Fuchs, R. L. *J. Agric. Food Chem.* **1990**, *38*, 1145-1152.
83. Tabashnik, B.; Finson, N.; Johnson, M. *J. Econ. Entomol.* **1992**, *85*, 2082-2087.
84. Zhang, J.; Wang, C.; Qin, J. *J. Invert. Pathol.* **2000**, *75*, 259-266.
85. Santos, M. O.; Adang, M. J.; All, J. N.; Boerma, H. R.; Parrott, W. A. *Mol Breed.* **1997**, *3*, 183-194.
86. Meade, T.; Hare, J. D. *J. Econ. Entomol.* **1995**, *88*, 1787-1794.
87. Walker, D. R.; Boerma, H. R.; All, J. N.; Parrott, W. A. *Mol. Breed.* 2001, *in press*.

Chapter 6

Genetic Engineering Crops for Improved Weed Management Traits

S. O. Duke[1], B. E. Scheffler[1], F. E. Dayan[1], and W. E. Dyer[2]

[1]NPURU, Agricultural Research Service, U.S. Department of Agriculture, P.O. Box 8048, University, MS 38677
[2]Plant Sciences Department, Montana State University, Bozeman, MT 59717

Crops may be genetically engineered for weed management purposes by making them more resistant to herbicides or by improving their ability to interfere with competing weeds. Transgenes for bromoxynil, glyphosate, and glufosinate resistance are found in commercially available crops. Other herbicide resistance genes are in development. Glyphosate-resistant crops have had a profound effect on weed management practices in North America, reducing the cost of weed management, while improving flexibility and efficacy. In general, transgenic, herbicide-resistant crops have reduced the environmental impact of weed management because the herbicides with which they are used are generally more environmentally benign and have increased the adoption of reduced-tillage agriculture. Crops could be given an advantage over weeds by making them more competitive or altering their capacity to produce phytotoxins (allelopathy). Strategies for producing allelopathic crops by biotechnology are relatively complex and usually involve multiple genes. One can choose to enhance production of allelochemicals already present in a crop or to impart the production of new compounds. The first strategy involves identification of the allelochemical(s), determination of their respective enzymes and the genes that encode them, and, the use of genetic engineering to enhance production of the compound(s). The latter strategy would alter existing biochemical pathways by inserting transgenes to produce new allelochemicals.

Weed management in agriculture has been dominated by the use of selective herbicides (herbicides that spare the crop while killing some of the important weed species in that crop) for the past 50 years. At this point in time, approximately 70 to 75% of the pesticides sold (by volume) in the U.S. are herbicides (*1*). Genetic engineering has provided alternatives to pesticide use in managing microbial and insect pests in crops (*e.g.*, *2, 3*), but the first transgenic crops designed for better weed management have been those which resist herbicides. This topic has been reviewed in two books (*4, 5*), and is the subject of numerous reviews (e.g., *6-10*). This chapter will provide a brief review of the area of herbicide-resistant crops (HRCs) produced by transgenic methods, and will discuss the possibility of genetically engineering crops to fight weeds without synthetic herbicide inputs.

Herbicide-Resistant Crops

Current Status

Since they were introduced in 1995, HRCs have been the largest segment of the transgenic crop market. Several crops that use herbicide resistance-imparting transgenes are now available in North America (Table I). Two resistance mechanisms have been used: modification of the herbicide target site to make it insensitive and enhancement of herbicide degradation by insertion of a transgene encoding a degradation enzyme.

Table I. Transgenic herbicide-resistant crops available in North America

Herbicides	Crop	Year introduced	Resistance mechanism
bromoxynil	cotton	1995	enhanced degradation
	canola	1999	enhanced degradation
glufosinate	maize	1997	altered target site
	canola	1997	altered target site
glyphosate	soybean	1996	altered target site
	canola	1997	altered target site & enhanced degradation
	cotton	1997	altered target site
	maize	1998	altered target site

The rapid adoption of this technology indicates that farmers find this trait to be very valuable. The rapid adoption of glyphosate-resistant cotton and soybeans is illustrated in Figure 1. Similarly, the use of glyphosate-resistant maize increased from 950,000 acres in 1998, when it was introduced, to 2.3

million acres in 1999. Other HRCs have been very useful tools for farmers with particular weed problems that were not well addressed by available weed management technologies. HRC use has been widely adopted in the U.S., Canada, and Argentina, while in some parts of the world there has been considerable public opposition to the use of this technology.

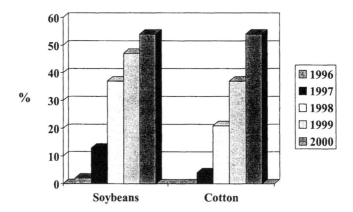

Figure 1. Adoption of glyphosate-resistant soybeans and cotton in the United States. Data are compiled from the USDA, Agricultural Marketing Service, USDA, NASS, and Monsanto Company by L. Gianessi and J. Carpenter of the National Center for Food and Agricultural Policy, Washington, DC.

Non-Selective Herbicides

Non-selective herbicides are those that kill virtually all plant species at a specified dose. Before the advent of genetic engineering, there was limited use of non-selective herbicides in agriculture. They were used with specialized equipment that prevented contact with the crop or when the objective was to kill all plants in an area, such as before planting or after harvest. Still, due to its many desirable traits (*11*), the non-selective herbicide glyphosate (*N*-(phosphonomethyl)glycine) has been used extensively. The advent of glyphosate-resistant crops greatly expanded the use of this herbicide in North America.

Glyphosate

Glyphosate is a highly effective, but environmentally and toxicologically safe, herbicide that inhibits a critical enzyme of the shikimate pathway, 5-

enolpyruvylshikimate-3-phosphate synthase (EPSPS). The shikimate pathway produces aromatic amino acids and a large number of secondary products, including lignins, flavonoids, and tannins. The enzyme does not exist in animals. Glyphosate is very mobile within the plant, with preferential transport to metabolic sinks such as meristematic tissues (*11*). It is relatively slow acting, so that it is transported throughout the plant before growing tissues are killed. For this reason, it is very effective in controlling perennial weeds in which subterranean tissues must be killed in order to prevent re-growth. Although some of the phytotoxicity of glyphosate is a result of reduced pools of aromatic amino acids, most of its the herbicidal effect appears to be caused by a general disruption of metabolic pathways through deregulation of the shikimate pathway (*12*).

Glyphosate-resistant crops required considerable research and development effort to produce (*13*). The greatest difficulty in obtaining a crop with sufficient resistance for commercial use was obtaining a glyphosate-resistant form of EPSPS that retained adequate catalytic efficiency to function well in the shikimate pathway. Simply amplifying gene expression of the glyphosate-susceptible form of the enzyme did not provide adequate levels of resistance for field use. Attempts to isolate a microbial gene encoding a C-P lyase that could degrade glyphosate in transgenic plants were unsuccessful. After exhaustive evaluation of both natural and mutant glyphosate-resistant forms of EPSPS, the naturally occurring CP4 EPSPS from *Agrobacterium* sp. strain CP4 was used to produce commercial glyphosate-resistant crops. Later, the gene encoding an enzyme that cleaves the C-N bond of glyphosate (glyphosate oxidase; *GOX*) was isolated from *E. coli*. The *GOX* gene has been used in combination with *CP4* in commercial glyphosate-resistant canola.

Neither the *CP4* nor the *GOX* gene imparts resistance to herbicides other than glyphosate. Thus, these genes are linked exclusively to one herbicide.

The rapid adoption of glyphosate-resistant crops is due to several factors. First, this technology greatly simplifies weed management (*14*). In many cases, it allows farmers to use only one herbicide, and only apply treatments after the weed problem develops. In those cases in which glyphosate is the only herbicide used, the farmer is less dependent on consultants for specialized recommendations for several herbicides that are sometimes applied at different times. Weed management with glyphosate-resistant crops generally requires less equipment, time, and energy than with selective herbicides. The efficacy of glyphosate in combination with glyphosate-resistant crops is generally very good. In many cases, it fills weed management gaps that existed with available selective herbicide (*15*). Furthermore, the economics of this approach, even with the "technology fee" added to the cost of the seed, are generally good. Most published economic analyses (*e.g., 16 17*) predict an economic advantage for glyphosate-resistant crops over conventional weed management; but, in a few cases, the economics are the same. The herbicide is no longer under patent protection and is being sold in numerous formulations and as several salts with differing cations. The declining cost of glyphosate due the expiration of its patent favors a continued economic advantage for glyphosate-resistant crop-based weed management.

The efficacy of any pest management strategy is never static, due in large part to pest species shifts and the evolution of resistance to management technologies. Although there are only two reported cases of evolved resistance to glyphosate (18-20), species that are more naturally resistant to glyphosate are likely to become problems in field situations in which glyphosate is used year after year. Already, in glyphosate-resistant soybeans in Iowa, a more glyphosate-resistant weed, common waterhemp (*Amaranthus rudis* L.), has become a problem where it was not a problem before (21). This and similar problems can be solved by rotating herbicides, mixing herbicides, and/or increasing the application rate of glyphosate.

Glufosinate

Glufosinate [2-amino-4-(hydroxymethylphosphinyl)butanoic acid] is the synthetic version of the microbially-produced natural product phosphinothricin (22). Like glyphosate, it is a non-selective herbicide that can be used to kill a broad spectrum of weed species. It is much faster acting than glyphosate, because it inhibits glutamine synthetase and thereby blocks photorespiration and photosynthesis. *Streptomyces hygroscopicus* produces an inactive tripeptide, bialaphos, which is converted to phosphinothricin in target organisms and apparently also in the producing organism. The producing microbe protects itself from autotoxicity with an enzyme that inactivates bialaphos, the bialaphos resistance (BAR) enzyme. *Streptomyces viridochromogenes* produces a similar enzyme, phosphinothricin acyl transferase (PAT), which inactivates phosphinothricin or glufosinate. These enzymes are encoded by the *bar* and *pat* genes, respectively, which can be used as transgenes to render crops resistant to glufosinate (22-24). In fact, glufosinate has been used extensively as a selection agent for transformants that contain the *pat* or *bar* gene linked to the transgene of interest. There are no analogs of glufosinate on the herbicide market, and these genes do not provide resistance to any other herbicides.

Almost every crop (major and minor) has now been transformed with the *bar* gene. However, at this time, this trait is commercially available in only two crops (Table I). Regulatory approval has been given for several glufosinate-resistant crops that are not yet commercially available. From a technical standpoint, this herbicide/HRC combination works quite well. But, the cost of this technology, compared to conventional weed management or glyphosate-resistant crop technology is not competitive in some cases.

Selective Herbicides

In agriculture, most of the herbicide market has been devoted to selective herbicides that kill problem weeds at doses that have minimal effects on the crop. The utility of many of these very effective herbicides is limited by the few crops with which they can be used and by the unacceptable crop damage that sometimes occurs in the crops for which they are intended. In most cases, the mechanism of crop resistance to selective herbicides is rapid metabolic degradation. Herbicide-resistance-imparting transgenes can be used to improve and expand the selectivity spectrum of these herbicides. There is currently only one commercial example of this approach, but others are under development.

Inhibitors of Acetolactate Synthase

The sulfonylurea and imidazolinone herbicides are very potent inhibitors of the acetolactate synthase (ALS), a key enzyme of branched chain amino acid synthesis (*25*). They represent a large segment of the herbicide market. Differential metabolic degradation is the mechanism of selectivity in crops in all cases, and specific sulfonylurea and imidazolinone herbicides have been designed for particular crops. However, certain weed species rapidly evolved resistance at the target site level to these herbicides (*26*). These weeds with a resistant form of ALS appear to pay little or no metabolic penalty for resistance. Thus, crops could be transformed with a resistant form of ALS to broaden the array of compatible ALS inhibitor herbicides and to reduce the potential for phytotoxicity on the crop.

A number of plant-derived, herbicide-resistant forms of ALS have been used as transgenes in the laboratory (*25*), and crops transformed with some of these have regulatory approval for field testing in the U.S. However, the ALS inhibitor-resistant crops produced by biotechnology that are commercially available have been produced by mutation and traditional breeding.

Bromoxynil

Bromoxynil (3,5-dibromo-4-hydroxybenzonitrile) is an inhibitor of photosystem II of photosynthesis. It is not a widely used herbicide. A microbe with a nitrilase that rapidly degrades bromoxynil was found in a bromoxynil-contaminated area. The gene encoding this enzyme was isolated and has been used to impart bromoxynil resistance in transgenic crops (*27*). The gene does not impart resistance to other classes of PS II-inhibiting herbicides, thus linking the transgenic crop to a specific herbicide.

The first introduced commercial HRC was bromoxynil-resistant cotton. This product has been extremely valuable for specific, but not widespread, weed problems (*15*). Although bromoxynil-resistant cotton has not had the adoption rate of glyphosate-resistant cotton, its use has steadily increased to about 7 to 8% of the cotton acreage in the U.S. in 1999 and 2000. Bromoxynil-resistant canola became available to Canadian farmers in 1999, but has had only limited success.

Other Herbicides

Resistance to a large number of other selective herbicides has been achieved with transgenes (*4*), but most of these will never be commercially available for economic, environmental, toxicological, or other reasons. However, additional HRCs are being developed. For example, crops made resistant to inhibitors of protoporphyrinogen oxidase (Protox) are being developed by Novartis Chemical Co. (*28*). Protox is a key enzyme in the synthesis of chlorophyll and other porphyrin-based molecules. When inhibited *in vivo*, its product rather than its substrate accumulates at high levels through a complex sequence of events (*29*). At these levels, the enzyme product, protoporphyrin IX, is highly toxic in the presence of light and molecular oxygen, killing photosynthetic plants very quickly through the generation of singlet oxygen. Theoretically, there are several mechanisms by which plants could be genetically engineered to be resistant to Protox inhibitors (*30*). The mechanism chosen by Novartis is to introduce a resistant form of Protox. Whether this form of Protox is resistant to all commercial Protox-inhibiting herbicides is not public knowledge at this time.

Regulatory approval for field testing of transgenic crops made resistant to 2,4-D, isoxazoles, dalapon, chloroacetanilides, and cyanamide with transgenes has been issued in the U.S. Whether any of these products will be commercialized is not certain.

The Future of Herbicide-Resistant Crops

Companies will not market a product unless there is a clear economic reward. With a HRC, the ideal situation is production of transgenic crops that are resistant only to an excellent, reasonably inexpensive, non-selective herbicide to which there is an economic link. To some extent, this has been the case with glyphosate- and glufosinate-resistant crops. However, the market niche has not been ideal for glufosinate resistance in some crops. We are aware of no other opportunities like these in development.

The future for HRCs that are resistant to selective herbicides is less certain. Selective herbicides already exist for all major crops. Thus, a crop that is genetically engineered to be resistant to yet another selective herbicide must fulfill a weed management need that is unmet, such as those use niches filled by bromoxynil-resistant crops. Most selective herbicides belong to herbicide classes represented by several commercial analogs, and thus most resistance transgenes are likely to provide resistance to all members of the herbicide class. The economics of profiting from a HRC tied to selective herbicides hinges on several factors, including: the cost of producing and developing the transgenic crop; whether or not there are economic links to manufacturers of the members of the herbicide class; and the degree of need for the product. Apparently, this equation has not produced positive results for several HRCs with resistance to selective herbicides.

Lastly, public opinion may play a critical role in the future use of HRCs. In a world economy, if a significant sector of a commodity market rejects transgenic crops, adoption will be crippled.

Crops that Interfere with Weeds

Plants can interfere with each other through competition for resources or by production of phytotoxins (allelopathy). Theoretically, either or both of these traits could be enhanced through genetic engineering in order to improve weed management.

Competition

Any efforts to impart the traits of faster growth, stress resistance (biological or physical), or more efficient and rapid utilization of light, water, and soil nutrients could give the crop an advantage over competing weeds. Selection and breeding for competitive ability have never been emphasized in modern plant breeding programs. However, in recent years it has become possible to consider transgenic approaches to enhance crop competitive ability due to our improved understanding of its underlying genetic mechanisms. Even though competitiveness is clearly a quantitative trait, significant progress has been made in identifying key components of the pathways controlling a plant's ability to compete for water, nutrients, and light. For example, plants have mechanisms that detect the proximity of neighboring plants and trigger anticipatory shade avoidance responses. This suite of adjustments in shoot morphology and physiology has been termed 'foraging for light' (*31*) and is controlled by several photoreceptors including phyB, a member of the phytochrome family of photoreceptors. Robson et al. (*32*) showed that this response could be functionally antagonized by overexpressing another phytochrome molecule (phyA) in transgenic tobacco plants. At high densities in the field, the transgenic plants showed significant changes in plant architecture and enhanced allocation of assimilates to leaves, modifications that result in net improvements in harvest index. These results suggest that further, and perhaps more subtle, manipulations of light response pathways could be used to modify a crop plant's growth habit and confer competitive advantages over weeds through niche occupation or light interception.

Crops often compete with weeds for scarce moisture and nutrients under stressful conditions. Recent work in plant stress physiology has identified a number of pathways and regulatory genes that enable some species to withstand extreme environmental conditions. Use of these genes in transgenic crop applications has been proposed (33, 34) and some promising results have already been obtained. For example, transgenic plants expressing the DREB1A transcription factor, which is known to induce several stress tolerance genes, showed enhanced tolerance to drought, freezing, and salt stress conditions (35). Significantly, these results show that improvements in tolerance to several common environmental stresses can be achieved through transfer of a single gene. The regulation of plant growth and development by phytohormones is also a promising target for biotechnology (36). Transgenic poplar trees overexpressing gibberellin biosynthetic genes showed elevated growth rates and biomass accumulation, compared to controls (37). Manipulation of plant hormones in order to increase growth rates during seedling growth, when competition with weeds is most critical, could be effective in improving competitive traits of crops.

The alteration of crop competitiveness through biotechnology is in its infancy. However, the few examples published to date suggest that there is incredible potential for modifying growth habit, nutrient uptake efficiency, resource allocation patterns, and other relevant traits of crop plants that make them more competitive with weeds. These traits have not been evaluated for their influence on competitive ability of crops with weeds. As our knowledge improves, it may also be possible to enhance beneficial relationships between crop plants and root colonizing bacteria, endophytes, or mycorrhizae. It is important to note that only very subtle improvements in some of these traits may be required in order to give crops a competitive advantage over their weedy neighbors.

Allelopathy

The ability of a crop to gain an advantage over weeds by producing phytotoxins is an approach to weed control that has fascinated scientists since the early 20th century. Although germplasms for allelopathy traits have been well established in several crops such as rice (38), barley (39), cucumber (40), sorghum (41), and wheat (42), plant breeders and geneticists have thus far not produced commercial varieties with an allelopathy trait. Genetic engineering offers tools that may allow for the production of allelopathic varieties of crops that would allow farmers to reduce their dependence on synthetic herbicides. A comparison of the characteristics of weed management using synthetic herbicides or through allelopathy is provided in Table II.

Enhancing existing allelopathic traits or imparting new ones into crops with transgenes will, in most cases, be a much more daunting task than producing a HRC. Effective allelopathy in a crop situation will involve synthesis of the allelochemical(s) in root tissue at sufficient levels and exudation of the compound into the soil (43). Each process will probably involve multiple gene products.

Table II. Comparison of weed management by allelopathy and by synthetic herbicides

Synthetic herbicides	Allelopathy
High input cost	Low input cost
More weather dependent	Less weather dependent
Highly effective	Less effective
Intermittent	Continuous
Less environmentally benign	More environmentally friendly

When using the existing allelopathic potential of a crop, the allelochemical that is to be manipulated must first be determined. The compounds contributing to allelopathy are not well defined for many crops such as rice (38). However, in sorghum, a highly potent root-exuded phytotoxin called sorgoleone has been well-characterized (44). Once an allelochemical has been chosen for enhancement, there are two ways to determine which genes would be good candidates to increase its production.

A very laborious approach is to determine the biochemical pathway leading to production of the compound, isolate key enzymes of the pathway, and then work back to the genes from the enzyme amino acid sequence (45). The alternative is to determine what genes are involved in the allelopathic trait and then determine the gene functions. This strategy can also be quite time-consuming. The latter approach can be accomplished either by creating mutants that have modified allelopathic traits or by differential screening of expressed genes of non-producing and producing tissues. After isolating putative genes, proof that a gene encodes an enzyme in expression of an allelopathic trait (a biosynthetic enzyme or a regulatory gene) must be made. The simplest method is to identify the gene by its homology to a known gene in a database, but this does not guarantee proof of function. If sequence homology is not found, the gene can be expressed in a microbe and the activity of the resulting protein determined. However, there are many factors that make this approach unsure. Another alternative is to overexpress or block expression of the gene (using anti-sense mRNA technology) in transgenic plants. The resulting effects on the allelopathic trait should offer clues as to the gene function.

Transgenic crops that synthesize novel compounds to resist insect and pathogen attack have been successfully produced. If the substrate for an enzyme that produces a phytotoxin is already in a metabolic pathway of a plant, engineering allelopathy could be relatively simple. However, autotoxicity (discussed below) could be a problem that would have to be dealt with. An example of what might be done is with crops engineered to be resistant to phosphinothricin or glufosinate (discussed above). These crops are already resistant to phosphinothricin, and introduction of the microbial genes for production of the phosphinothricin precursor, bialaphos, by the plant is possible without unacceptable autotoxicity. These genes are available (46).

After isolation and identification of a candidate gene for manipulation of crop allelopathy, the crop must be transformed. To be most effective, the transgene should be controlled by a promoter that only allows its expression in root tissue, or even better, only in root hairs. Root-specific promoters are available (e.g., 47), but root hair specific promoters are not yet available in the public sector. Further manipulations would be needed to ensure that the allelochemical is secreted into the rhizosphere.

Even if key genes are isolated and expressed in specific tissues in transformed crops, there are still several potential pitfalls for this strategy. First, manipulation of a particular pathway could cause unacceptable metabolic imbalances or imbalances in resource partitioning. An example of the former is that of Canel et al. (48) in which overexpression of tryptophan carboxylase in Catharanthus roseus led to greatly reduced growth because of disruption of metabolism by depleted tryptophan pools. Even if growth is not severely disrupted, partitioning of resources into large amounts of a secondary product could reduce yield. Gershenzon (49) has determined that many secondary products such as allelochemicals are metabolically costly. These problems would be minimized if the allelochemical has a high unit activity as a phytotoxin, and thus requires relatively little diversion of metabolites from other pathways for adequate concentrations to be produced. Tissue-specific expression should also reduce the probability of this problem.

Autotoxicity is another problem that could seriously limit the use of allelochemicals in crops. If this problem is of sufficient magnitude, engineering resistance could alleviate it. Approaches to this problem are discussed in detail by Scheffler et al. (43).

The probability of weeds rapidly evolving resistance to allelochemicals is unknown. This is a complex question that cannot be easily answered and for which there is almost no information available. A recent study suggests that the reason that Centaurea diffusa is much more successful in North America than in its native Eurasia, is that the plant species in its geographic origin have evolved resistance to allelochemicals that it produces (50). The probability of resistance occurring can be estimated by treating a large population of mutagenized seed with the minimum level of the phytotoxin to cause lethality (e.g., *13*). However, reliance on the response of a single species can be misleading (*10*).

Lastly, there could be a problem with effects of the allelochemical on non-target organisms, including humans. Natural compounds usually have shorter environmental half-lives than synthetic herbicides, but many natural compounds are toxic to a variety of organisms. This aspect would have to be evaluated before release of an allelopathic crop variety.

Risks and Benefits

The benefits and risks of the technologies discussed above would vary with the specific crop and trait. Even for a specific transgenic crop, this assessment might vary dramatically from one geographic area to another. Furthermore, risks and benefits must be evaluated in the context of the risks and benefits of current agricultural practices. Thus, generalizations should be made with the caveat that they may not apply to every situation. Table III provides some generalizations that are discussed in much more detail in other publications (*4-7, 43*).

Table III. Summary of some of the potential risks and benefits of transgenic herbicide-resistant and allelopathic crops

Risks	Benefits
Herbicide-resistant crops	
Continued reliance on herbicides	Reduced use of less benign herbicides
Transgene flow to weeds	Increased adoption of reduced tillage
Increased non-target problems	Increased weed management flexibility
Lack of consumer acceptance	Reduced weed management costs
Increased selection pressure for evolved herbicide resistance	
Improved crop interference	
Transgene flow to weeds	Decreased reliance on herbicides
Lack of consumer acceptance	Increased adoption of reduced tillage
Crop becoming an invasive weed	Increased weed management flexibility
	Reduced input and cost for farmer

Enough time has passed since the introduction of HRCs to conclude that the relatively benign herbicides for which they are designed will reduce use of some environmentally and toxicologically suspect herbicides. Nevertheless, such crops will perpetuate the reliance on synthetic herbicides for weed management.

We appear to be far from genetically engineering crops to effectively combat weeds through improved competition and/or allelopathy. If successful, this technology would greatly reduce the reliance on synthetic herbicides. The biggest risk of either genetic engineering approach appears to be that of gene flow to weedy relatives. In the case of herbicide resistance, the transgene should have no effect in natural ecosystems in which the herbicide is not used. However, traits that produce a more competitive or more allelopathic plant have the potential to alter a natural ecosystem. Reproductive barriers should considered for use with those traits that appear to present such a threat.

Acknowledgments

We thank Leonard Gianessi and Janet Carpenter for providing the data in Figure 1.

References

1. Anonymous. *Agrow World Crop Protection News.* **1998.** *304,* 16.
2. Beachy, R. N.; Bendahmane, M. In *Emerging Technologies for Integrated Pest Management*; Kennedy, G. G.; Sutton, T. B., Eds.; Amer. Phytopath. Soc. Press: St. Paul, MN, 2000; pp. 101-107.
3. Fitt, G. P.; Wilson, L. J. In *Emerging Technologies for Integrated Pest Management*; Kennedy, G. G.; Sutton, T. B., Eds.; Amer. Phytopath. Soc. Press: St. Paul, MN, 2000; pp. 108-125.
4. *Herbicide-Resistant Crops*; Duke, S. O., Ed.; CRC Press, Boca Raton, FL, 1996; 420 pp.
5. *Herbicide-Resistant Crops and Pastures in Australian Farming Systems*; McLean, G. D.; Evans, G., Eds. ; Bureau of Resource Sciences, Canberra, Australia, 1995; 292 pp.
6. Hess, F. D.; Duke S. O. In *Emerging Technologies for Integrated Pest Management*; Kennedy, G. G.; Sutton, T. B., Eds.; Amer. Phytopath. Soc. Press: St. Paul, MN, 2000; pp. 126-140.
7. Duke, S. O. *J. Weed Sci. Technol. (Zasso-Kenkyu,* Japan). **1998.** *43,* 94-100.
8. Dekker, J.; Duke, S. O. *Adv. Agron.* **1995.** *54,* 69-116.
9. Gressel., J. *Plant Breeding Rev.* **1993.** *11,* 155-198.
10. Gressel, J. *Transgenic Res.* **2000.** *9,*355-382.
11. Duke, S. O. In. *Herbicides: Chemistry, Degradation, and Mode of Action, Vol. 3*; Kearney, P. C.; Kaufman, D R., Eds.; Dekker, New York, 1998; pp. 1-70.

12. Siehl, D. L. In *Herbicide Activity: Toxicology, Biochemistry and Molecular Biology*; Roe, R. M.; Burton, J. D.; Kuhr, R. J., Eds.; IOS Press, Amsterdam, 1997; pp. 37-67.

13. Padgette, S. R.; Re, D. B.; Barry, G. G.; Eichholtz, D. E.; Delannay, X.; Fuchs, R. L.; Kishore, G. M.; Fraley, R. T. In *Herbicide-Resistant Crops*; Duke, S. O., Ed.; CRC Press, Boca Raton, FL, 1996; pp. 53-84.

14. Barnes, R. L. *Pest Manag. Sci.* **2000**. *56*, 580-583.

15. Baldwin, F. L. *Pest Manag. Sci.* **2000**. *56*, 584-585.

16. Reddy, K. N.; Whiting, K. *Weed Technol.* **2000**. *14*,104-211.

17. Johnson, W. G.; Bradley, P. R.; Hart, S. E.; Buesinger, M. L.; Massey, R. E. *Weed Technol.* **2000**. *14*, 57-65.

18. Powles, S. B.; Lorraine-Colwill, D. F.; Dellow, J. J.; Preston, C. *Weed Sci.* **1998**. *46*,604-607.

19. Dill, G.; Baerson, S.; Casagrande, L.; Feng, Y.; Brinker, R.; Reynolds, T.; Taylor, N.; Rodriguez, D.; Teng, Y. *Proc. 3rd Inter. Weed Sci. Cong.* **2000**. p. 150.

20. Lee, L. J.; Ngim, J. *Pest Management Sci.* **2000**. *56*,336-339.

21. Owen, M. D. K. *Proc. Brighton Crop Protect. Cong.* **1997**. *3*,955-963.

22. Lydon, J.; Duke, S. O. In *Amino Acids*; Singh, B. J., Ed.; Dekker, New York, 1999; pp.

23. Vasil, I. K. In Herbicide-Resistant Crops; Duke, S. O., Ed.; CRC Press, Boca Raton, FL, 1996; pp. 85-91.

24. Müllner, H.; Eckes, P., Donn, G. *Amer. Chem. Soc. Symp. Ser.* **1993**. 524,38-47.

25. Saari, L. L.; Mauvais, C. J. In *Herbicide-Resistant Crops*; Duke, S. O., Ed.; CRC Press, Boca Raton, FL, 1996; pp. 127-142.

26. Saari, L. L.; Cotterman, J. C.; Thill, D. C. In *Herbicide Resistance in Plants*; Powles, S. B.; Holtum, J. A. M., Eds.; Lewis Pub., Boca Raton, FL, 1994; pp. 83-139.

27. Stalker, D. M.; Kiser, J. A.; Baldwin, G.; Coulombe, B.; Houck, C. M. In *Herbicide-Resistant Crops*; Duke, S. O., Ed.; CRC Press, Boca Raton, FL, 1996; pp. 93-105.

28. Volrath, S. L.; Johnson, M. A.; Ward, E. R.; Heifetz, P. B. U.S. Patent 6,084,155, 2000.

29. Dayan, F. E.; Duke, S. O. In *Herbicide Activity: Toxicology, Biochemistry and Molecular Biology*; Roe, R. M.; Burton, J. D.; Kuhr, R. J., Eds.; IOS Press, Amsterdam, 1997; pp. 11-35.

30. Duke, S. O.; Dayan, F. E.; Yamamoto, M.; Duke, M. V.; Reddy, K. N.; Lee, H. J.; Jacobs, N. J.; Jacobs, J. M. *Proc. 2nd Internat. Weed Control Cong., Copenhagen.* **1996**. *3*:775-780.

31. Ballaré C. L.; Scopel, A. L. ; Sánchez, R. A. *Plant, Cell Environ.* **1997**. *20*,820-825.

32. Robson, P. R. H.; McCormac, A. C.; Irvine, A. S.; Smith, H. *Nature Biotechnol.* **1996**. *14*,995-998.
33. Bajaj, S.; Targolli, J.; Liu, L. F.; Ho, T. H.-D.; Wu, R. *Molec. Breeding* **1999**. *5*:493-503.
34. Van Breusegem, F.; Van Montagu, M.; Inze, D. *Outlook Agric.* **1998**. *27*(2),115-124.
35. Kasuga, M.; Liu, Q.; Miura, S.; Yamaguchi, S.K.; Shinozaki, K. *Nature Biotechnol.* **1999**. *17*(3):287-291.
36. Hedden, P.; Phillips, A. L. *Current Opinions Biotechnol.* **2000**. *11*(2), 130-137.
37. Eriksson, M. E.; Israelsson, M.; Olsson, O.; Mortiz, T. *Nature Biotechnol.* **2000**. *18*,784-788.
38. Olofsdotter, M.; Wang, D.; Navarez, D. In *Recent Advances In Allelopathy. Vol. I. A Science for the Future*; Macias, F. A.; Galindo, J. C. G.; Molinillo, J. M. G.; Cutler, H. G., Eds.; Univ. Cadiz Press, Cadiz, Spain, 1999; pp. 383-390.
39. Lovett, J. V.; Hoult, A. H. C. *Amer. Chem. Soc. Symp. Ser.* **1995**. *582*,170-183.
40. Putnam, A. R.; Duke, W. B. *Science* **1974**. *185*,370-372.
41. Nimbal, C. I., Yerkes, C. N.; Weston, L. A.; Weller, S. C. *J. Agric. Food Chem.* **1996**. *44*,1343-1347.
42. Wu, H.; Pratley, H.; Lemerle, D.; Haig, T. *Weed Res.* **1999**. *39*,171-180.
43. Scheffler, B. E.; Duke, S. O.; Dayan, F. E.; Ota, E., *Rec. Adv. Phytochem.* **2001** (*in press*).
44. Rimando, A. M.; Dayan, F. E.; Czarnota, M. A.; Weston, L. A.; Duke, S. O. *J. Nat. Prod.* **1998**. *61*,927-930.
45. Canel, C. *Acta Hort.* **1999**. *500*,51-57.
46. Lydon, J. *Plant Growth Regul. Soc. Amer. Quarterly* **1996**. *24*,111-139.
47. Yamamoto, Y. T.; Taylor, C. G.; Acedo, G. N.; Cheng, C. L.; Conkling, M. A. *Plant Cell* **1991**. *3*,371-382.
48. Canel, C.; Lopes-Cardoso, M. I.; Whitmer, S.; van der Fits, L.; Pasquali, G.; van der Heijden, R.; Hoge, H.; Verpoorte, R. *Planta* **1998**. *205*,414-419.
49. Gershenzon, J. *J. Chem. Ecol.* **1994**. *20*,1281-1328.
50. Callaway, R. M.; Aschehoug, E. T. *Science* **2000**. *290*,521-523.

Chapter 7

Environmentally Friendly Approaches in Biotechnology: Engineering the Chloroplast Genome to Confer Stress Tolerance

H. Daniell

Department of Molecular Biology and Microbiology and Center
for Discovery of Drugs and Diagnostics, University of Central Florida,
12722 Research Parkway, Orlando, FL 32826–3227

Chloroplast genetic engineering is emerging as an alternative new technology to overcome some of the potential environmental concerns of nuclear genetic engineering (reviewed in ref 1). One commonly perceived environmental concern is the escape of foreign gene through pollen or seed dispersal from transgenic crop plants to their weedy relatives creating super weeds or causing genetic pollution among other crops (2). High rates of such gene flow from crops to wild relatives (as high as 38% in sunflower and 50% for strawberries) are certainly a serious concern. Keeler et al. (3) have summarized valuable data on the weedy wild relatives of sixty important crop plants and potential hybridization between crops and wild relatives. Among sixty crops, only eleven do not have congeners (members of the same genus) and the rest of the crops have wild relatives somewhere in the world. In addition, genetic pollution among crops has resulted in several lawsuits and shrunk the European market for organic produce from Canada from 83 tons in 1994–1995 to 20 tons in 1997–1998 (4). For example, a canola farmer in Canada cultivated a glyphosate (Round-up) resistant cultivar (Quest) and a glufosinate (Liberty)-resistant cultivar (Innovator) 30 meters away across an intervening road that exceeds the standard buffer zone of 6 meters. Two applications of Round-up herbicide in 1998 to the field sown with glufosinate resistant cultivar killed all the weeds but revealed glyphosate resistant canola in the field sown with other cultivars. This population was thickest near the road, where airborne dispersal of pollen from glyphosate resistant canola could occur. Meanwhile, a Canadian farmer is being sued by Monsanto for possessing and growing glyphosate resistant canola without a license, however, the farmer claims that his crops were contaminated by resistant genes via wind or bee pollination. Because of all these concerns, Canadian National Farmers Union is lobbying the Canadian

Federal Government to legislate industry compensation for unintended genetic alteration of crops (4). Several major food corporations, including ADM have required segregation of native crops from those polluted with transgenes. Two legislations have been submitted in the U.S. to protect organic farmers whose crops inadvertently contain transgenes via pollen drift (5). Maternal inheritance of foreign genes through chloroplast genetic engineering is highly desirable in such instances where there is potential for out-cross among crops or between crops and weeds (6-8).

Yet another concern in the use of nuclear transgenic crops expressing the *Bacillus thuringiensis (Bt) toxins* is the sub-optimal production of toxins resulting in increased risk of pests developing Bt resistance. Plant-specific recommendations to reduce Bt resistance development include increasing Bt expression levels (high dose strategy), expressing multiple toxins (gene pyramiding), or expressing the protein only in tissues highly sensitive to damage (tissue specific expression). All three approaches are attainable through chloroplast transformation. For example, hyper-expression of several thousand copies of a novel B.t. gene via chloroplast genetic engineering resulted in 100% mortality of insects that are up to 40,000-fold resistant to other B.t. proteins (9). Another hotly debated environmental concern expressed recently is the toxicity of transgenic pollen to non-target insects, such as the Monarch butterflies (10,11). In contrast to this observation, expression of very high levels of insecticidal proteins in leaves (up to 50% of total soluble protein), did not result in accumulation of toxic proteins in pollen of chloroplast transgenic plants (12). Chloroplast gene expression also results in tissue specificity, occurring predominantly where functional plastids are present. This may be important in engineering insect resistant plants wherein most worms predominantly feed on leaves where plastids are abundantly present, thereby consuming the highest level of the insecticidal protein. If desired, regulatory signals specific for non-green plastids should be used to engineer insect resistance in fruits or tubers.

Another remarkable feature of chloroplast genetic engineering is the observation of exceptionally large accumulation of foreign proteins in transgenic plants, including more than 50% of CRY protein in total soluble protein, even in bleached old leaves (12, 13). Stable expression of a pharmaceutical protein in chloroplasts was first reported for GVGVP, a protein based polymer with varied medical applications (such as the prevention of post-surgical adhesions and scars, wound coverings, artificial pericardia, tissue reconstruction and programmed drug delivery, 14). Subsequently, expression of the human somatotropin via the tobacco chloroplast genome (15) to high levels (7% ot total soluble protein) was observed. It is well known that the level of foreign gene expression is not adequate for commercial feasibility of several pharmaceutical proteins when expressed via the nuclear genome; levels of expression of pharmaceutical proteins vary over three orders of magnitude, 0.001 to 1% of total soluble protein (16). Therefore, it is wise to exploit this major advantage by engineering foreign genes via the chloroplast genome instead of the nuclear genome. Hyper-expression at the site of infection is also highly desirable to control invasion of pathogens in transgenic plants. Because of the concentration dependent action of the anti-microbial peptide (MSI 99) in controlling bacterial and fungal infection, we expressed it via the chloroplast genome to accomplish 100% mortality of bacteria and fungi at the point of infection (17). Chloroplast transformation utilizes two flanking sequences that, through homologous recombination, insert foreign DNA into the spacer region between the functional genes of the chloroplast genome, thus targeting the foreign genes to a precise location. This eliminates the "position effect" and gene silencing observed in nuclear transgenic plants (18,19).

In plant and animal cells, nuclear mRNAs are translated monocistronically. This poses a serious problem when engineering multiple genes in plants (1). Therefore, in order to express the polyhydroxybutyrate polymer or Guy's 13 antibody, single genes were first introduced into individual transgenic plants, then these plants were back-crossed to reconstitute the entire pathway or the complete protein (20,21). Similarly, in a seven yearlong effort, Ye et al (22) recently introduced a set of three genes for a short biosynthetic pathway that resulted in β-carotene expression in rice. In contrast, most chloroplast genes of higher plants are co-transcribed and co-translated (1). Multiple steps of chloroplast mRNA processing are involved in the formation of mature mRNAs. Expression of polycistrons via the chloroplast genome provides a unique opportunity to express entire pathways in a single transformation event. The first example of a bacterial operon expression in transgenic plants, engineered via the chloroplast genome (12) will be discussed in this review.

There have been several efforts to generate various stress resistant transgenic plants by introducing gene(s) responsible for trehalose biosynthesis, regulation or degradation (23-25). When trehalose accumulation was increased in transgenic tobacco plants by over-expression of the yeast *TPS1*, trehalose accumulation resulted in the loss of apical dominance, stunted growth, lancet shaped leaves and some sterility. Altered phenotype was always correlated with drought tolerance; plants showing severe morphological alterations had the highest tolerance under stress conditions. Several toxic compounds expressed in transgenic plants have been compartmentalized in chloroplasts even through no targeting sequence was provided (26,27), indicating that this organelle could be used as a repository like the vacuole. Also, osmoprotectants are known to accumulate inside chloroplasts under stress conditions (29). Inhibition of trehalase activity in the cytosol is known to enhance trehalose accumulation in plants (24). Therefore, trehalose accumulation in chloroplasts may be protected from trehalase activity in the cytosol, if trehalase was absent in the chloroplast. In order to minimize the pleiotropic effects observed in the nuclear transgenic plants accumulating trehalose, a recent study attempted to compartmentalize trehalose accumulation within chloroplasts. An example of trehalose accumulation inside chloroplasts and resultant drought tolerant phenotypes (28) is discussed in this review.

Most transformation techniques co-introduce a gene that confers antibiotic resistance, along with the gene of interest to impart a desired trait. Regenerating transformed cells in antibiotic containing growth media permits selection of only those cells that have incorporated the foreign genes. Once transgenic plants are regenerated, antibiotic resistance genes serve no useful purpose but they continue to produce their gene products. One among the primary concerns of genetically modified (GM) crops is the presence of clinically important antibiotic resistance gene products in transgenic plants that could inactivate oral doses of the antibiotic (reviewed in 30; 31). Alternatively, the antibiotic resistant genes could be transferred to pathogenic microbes in the gastrointestinal tract or soil rendering them resistant to treatment with such antibiotics. Antibiotic resistant bacteria are one of the major challenges of modern medicine. In Germany, GM crops containing antibiotic resistant genes have been banned from release (32). However, several approaches are currently available to eliminate antibiotic resistance genes from nuclear transgenic crops (30). An example of marker-free chloroplast genetic engineering will be discussed in this review.

Engineering herbicide resistance via the chloroplast genome

Selective herbicides are routinely applied to control weeds that would otherwise compete for available nutrients, space and light, thereby reducing crop yield. For example, glyphosate is a potent, broad-spectrum herbicide, which is highly effective against a majority of grasses and broad leaf weeds. Glyphosate works by competitive inhibition of the enzyme 5-enol-pyruvyl shikimate-3-phosphate synthase (EPSPS) of the aromatic amino acid biosynthetic pathway. Synthesis of EPSP from shikimate 3-phosphate and inorganic phosphate is catalyzed by EPSPS. This particular reaction occurs only in plants and microorganisms, which explains the non-toxicity of glyphosate to other living forms. Use of glyphosate is environmentally safe as it is inactivated rapidly in soil, has minimum soil mobility, and degrades to natural products, with little toxicity to non-plant life forms. However, glyphosate lacks selectivity and does not distinguish crops from weeds, thereby restricting its use. EPSPS based glyphosate resistance has been genetically engineered by the overproduction of the wild type petunia EPSPS (33) or by the expression of a mutant gene (*aro*A) encoding glyphosate resistant EPSPS (34). In all of the aforementioned examples, without exception, herbicide resistant genes have been introduced into the nuclear genome.

One common environmental concern is the escape of a foreign gene through pollen or seed dispersal, thereby creating super weeds or causing genetic pollution among other crops. Escape of herbicide resistance genes to wild relatives occurs predominantly via dispersal of viable pollen. Keeler et al. (3) focus on the role of gene flow to weedy wild relatives as a potential problem because in their opinion "this is a far greater concern than any other mode of escape of transgenes." These authors further point out that "transgenes can only reach weed populations if carried to weeds on viable pollen; if the crop produces no pollen or non-viable pollen, there will be no gene flow." The potential for gene flow via pollen depends on several factors including the amount of pollen produced, longevity of pollen, dispersal of pollen (via wind, animal), plant/weed density, dormancy/rehydration of pollen, survival of pollen from toxic substances secreted by pollinators and distance between crops and weeds. Keeler et al. (3) point out that it is impractical to prevent out-cross between weeds and wind pollinated crops because of the large pollen clouds produced and distance traveled by viable pollen.

However, it is possible under exceptional circumstances, for the herbicide-resistant crop plants to be fertilized by pollen from wild relatives and serve as female parent to produce hybrid seed. If this happens, the hybrid seed may germinate and establish a resistant population. However, for this to happen, the herbicide resistant crop that served as the female parent must escape harvesting and the hybrid seeds must survive to germinate, grow and reproduce. Alternatively, dispersal of seeds from transgenic plants may occur among weedy relatives, during harvest, transportation, planting and harvest. This can give rise to mixed populations. Introgressive hybridization could result in super weeds. This again would depend on the persistence of the crop among weeds and probability of forming mixed strands.

Genetic containment methods include apomixis, incompatible genomes, transgenetic mitigation, control of seed dormancy, seed ripening or shattering, suicide genes, infertility barriers, male sterility and maternal inheritance (35). The latter two have been experimentally tested. Anther, the male reproductive organ, is composed of several cell and tissue types and contains anther specific mRNA's (36). Anther produces pollen grains that contain sperm cells. A specialized anther tissue called the tapetum plays an important role in the formation of pollen. The tapetum generally surrounds the pollen sac in early development and is not present as an organized tissue in the mature anther. The tapetum synthesizes a number of proteins that aid in pollen development or become components of pollen. Many male sterility mutations interfere with the tapetal cell differentiation and/or function, indicating that this tissue is essential for the production of functional pollen. Mariani et al. (36) have shown that the 5'-region of a tobacco tapetum-specific gene (TA29) can activate the expression of β-glucuronidase and riboculease genes (RNase T1 and barnase) within the tapetal cells of transgenic tobacco and oil seed rape plants. Expression of RNase genes selectively destroyed the tapetum during anther development, prevented pollen formation and produced male sterile plants. This approach could be used to contain out-cross of transgene with other crops or weeds. However, male sterility is possible only in crops where the product is not a seed or fruit requiring fertilization (like lettuce, carrot or cabbage).

Scott and Wilkinson (7) have recently analyzed several factors that would influence the transgene movement of chloroplast genes from crops to wild relatives under natural conditions. They studied the mode of inheritance of plastids, incidence of sympatry to quantify opportunities for forming mixed populations and persistence of crops outside agriculture limits for introgression. They studied plastid inheritance in natural hybrids collected from two wild B. rapa populations growing next to oilseed rape along 34 km of the Thames River and assessed the persistence of 18 feral oil seed rape populations over a period of three years. These studies concluded that there would be no pollen-mediated transgene movement from oilseed rape. A low incidence of sympatry (0.6-0.7%) between the crop and weed species occurred; however, mixed strands showed a strong tendency towards rapid decline in plant number, seed return and ultimately extinction within three years. Thus, they concluded that gene flow will be rare if plants are genetically engineered via the chloroplast genome.

The prevalent pattern of plastid inheritance found in the majority of angiosperms is uniparental maternal and chloroplast genomes are maternally inherited for most of the crops. However, there are always exceptions to most observations and maternal inheritance of chloroplast genomes is certainly not without exception. It is known that in pines (gymnosperms) and a few flowering plants (like alfalfa) plastids are transmitted in a biparental mode. Paternal transmission of plastids in tobacco has been reported, but with provisos. In transmission of paternal chloroplasts in tobacco, authors mention that there is occasional (0.07-2.5%) paternal transmission in a species typically exhibiting strict maternal inheritance (see ref 37).

Maternal inheritance of a herbicide resistance gene and prevention of escape via pollen has been successfully demonstrated recently (6). Engineering foreign genes through chloroplast genomes (which are maternally inherited in most of the crops) is a practical solution to this problem. In addition, the target enzymes or proteins for most herbicides (of the amino acid / fatty acid biosynthetic pathways or photosynthesis) are compartmentalized within the chloroplast. Because the transcriptional and translational machinery of the chloroplast is prokaryotic in nature, herbicide resistant genes of bacterial origin can be expressed at extraordinarily high levels in chloroplasts. The first report of engineering herbicide resistance via chloroplast genome, in order to overcome out-cross and gene pollution concerns is discussed below.

The chloroplast vector pZS-RD-EPSPS contained the 16S rRNA promoter (Prrn) driving the *aadA* (aminoglycoside adenyl transferase) and EPSPS genes with the *psbA* 3' region (the terminator from a gene coding for photosystem II reaction center components) from the tobacco chloroplast genome. This construct integrated the EPSPS and *aadA* genes into the spacer region between the *rbcL* (the gene for the large subunit of RuBisCO) and *orf*512 genes (code for the *accD* gene) of the tobacco chloroplast genome. This vector is useful to integrate foreign genes specifically into the tobacco chloroplast genome; this gene order is not conserved among other plant chloroplast genomes (38). On the other hand, the universal chloroplast expression and integration vector pSBL-RD-EPSPS can be used to transform chloroplast genomes of several other plant species because the flanking sequences are highly conserved among higher plants; the universal vector uses *trnA* and *trnI* genes (chloroplast transfer RNAs coding for Alanine and Isoleucine) from the inverted repeat region of the tobacco chloroplast genome as flanking sequences for homologous recombination.

Transgenic plants were obtained within 3-5 months after bombardment as described by Daniell (39,40). The integration of the aroA gene into the chloroplast was confirmed by PCR and Southern analyses. In addition, the high level of resistance to glyphosate observed, was confirmed by determination of the copy number of the aroA gene, in the transgenic plants. The copy number of the integrated gene was determined by establishing homoplasmy for the transgenic chloroplast genome. Tobacco Chloroplasts contain 5000~10,000 copies of their genome per cell. If only a fraction of the genomes were actually transformed, the copy number, by default, must be less than 10,000. By establishing that in the transgenics the EPSPS transformed genome was the only one present, one could establish that the copy number is 5000~10,000 per cell. This proved that only the transgenic chloroplast genome was present in the cell and there was no native, untransformed, chloroplast genome, without the EPSPS gene present. This established the homoplasmic nature of transformants, simultaneously providing an estimate of about 10,000 copies of the foreign EPSPS gene per cell. This explained the high levels of tolerance to glyphosate observed in transgenic plants.

Seeds collected from transgenic plants after the first self-cross were germinated in the presence of spectinomycin (resistance conferred by the aadA gene). All of the seeds germinated, remained green and grew normally. The 100% resistance to spectinomycin in all of the clones examined showed maternal inheritance of the introduced genes. A heteroplasmic condition would have given rise to variegated progeny on spectinomycin; lack of such variegated progeny also confirms homoplasmy as confirmed by Southern blot analysis. All of the untransformed seedlings were bleached and did not grow in the presence of spectinomycin. Lack of variation in chlorophyll pigmentation among the progeny also underscores the absence of position effect, an artifact of nuclear transformation.

Eighteen week old control and transgenic plants were sprayed with equal volumes of different concentrations (0.5 to 5 mM) of glyphosate. Untransformed control tobacco plants were extremely sensitive to glyphosate; they died within seven days even at 0.5 mM glyphosate. On the other hand, the chloroplast transgenic plants survived concentrations as high as 5mM glyphosate. These results are intriguing, considering the fact that the EPSPS gene from petunia used in these chloroplast vectors, has a low level of tolerance to glyphosate. Sensitivity to glyphosate by EPSPS should have been compensated by overproduction of the enzyme by thousands of copies of the EPSPS gene, present in each cell of the transgenic plants. Also, this is the first report of expressing a eukaryotic nuclear gene within the prokaryotic chloroplast compartment. It is well known that the codon preference is significantly different between the prokaryotic chloroplast compartment and the eukaryotic nuclear compartment. Ideally, a mutant aroA gene from a prokaryotic system (which does not bind glyphosate) should be expressed in the chloroplast compartment; such genes are now available and exhibit a thousand fold higher level of resistance to glyphosate than the petunia gene used in this investigation. In light of these observations, it is possible that integration of prokaryotic herbicide resistance genes into the chloroplast genome could result in incredibly high levels of resistance to herbicides while still maintaining the efficacy of biological containment.

Engineering insect resistance via the chloroplast genome
The use of commercial, nuclear transgenic crops expressing *Bacillus thuringiensis* (Bt) toxins has escalated in recent years due to their advantages over traditional chemical insecticides. However, in crops with several target pests, each with varying degrees of susceptibility to Bt (e.g. cotton), there is concern regarding the sub-optimal production of toxin, resulting in reduced efficacy and increased risk of Bt resistance. Additionally, reliance on a single (or similar) B.t. protein(s) for insect control increases the likelihood of B.t.-resistance development (41). Most current commercial transgenic plants that target lepidopteran pests contain either Cry1Ab (corn) or Cry1Ac (cotton) (42,43). Bt corn is targeted primarily against European corn borer although other pests such

as the corn earworm or cotton bollworm may be affected. B.t. cotton is targeted primarily against the tobacco budworm; however, other pests such as armyworms and cotton bollworm are economically damaging, but have only limited susceptibility to Cry1Ac.

Use of single Bt protein to control insects such as tobacco budworm and cotton bollworm could lead to relatively rapid Bt resistance development (44,45). Additionally, because Cry1Ab and Cry1Ac share over 90% protein homology, resistance to one Cry1A protein would most likely impart resistance to another Cry1A protein as has been observed in tobacco budworm (45,46). Nowhere is this more of a concern than with cotton bollworm/ corn ear worm which usually feeds on corn in the spring and early summer, then migrates over to cotton to complete several more generations (44). Clearly, different Bt proteins are needed in order to decrease the development of resistance. The primary strategy currently used to delay development of insect resistance to *Bt* plants is to provide refuges of host plants that do not produce B.t. toxins. However, a recent study (47) of a resistant strain of pink bollworm larvae on B.t. cotton shows developmental asynchrony - this favors assortative mating among resistant moths emerging from B.t. plants, and generates a disproportionately high number of homozygous resistant insects, accelerating the evolution of B.t. resistance.

Another environmental concern expressed recently is the toxicity of transgenic pollen to non-target insects, including Monarch butterflies (10) although this study has been criticized as being premature and incomplete (11). Toxic insecticidal protein was not observed in pollen of chloroplast transgenic plants, despite very high levels of insecticidal protein accumulation (up to 50% of total soluble protein) in transgenic leaves (12). Evolving levels of B.t.-resistance in insects should be dramatically reduced through genetic engineering of the chloroplast genome in transgenic plants. Therefore, the first report of chloroplast genetic engineering that demonstrated 100% mortality of B.t. resistant insects is discussed below.

The tobacco chloroplast expression vector described above was also used to introduce a novel B.t. coding sequence into the chloroplast genome. This class of Bt proteins, Cry2A, is toxic to many caterpillars, such as the European corn borer and tobacco budworm, is quite different in structure/function from the Cry1A proteins (resulting in less cross resistance). Cry2A proteins are about half the size of Cry1A proteins, and therefore should be expressed at higher levels. Tobacco leaves were bombarded with DNA-coated tungsten particles as described elsewhere (39,40). The positive clones were analyzed by PCR and Southern hybridization to confirm the site-specific integration of *cry*2Aa2, and to establish copy number as explained before. Insect bioassays resulted in the following observations (9). There was 100% mortality of tobacco budworm feeding on transgenic leaves and the leaf pieces were essentially intact, while the control leaf pieces were completely devoured. Similar results were observed with CryIAc (up to 40,000 fold resistant) and Cry2A (up to 400 fold resistant) insects. Bioassays also were conducted using insects that were reared on control leaves or artificial diet for 5 days (ca. 2nd-3rd instar), and then moved to transgenic leaves. Even these older larvae that are more tolerant than neonates, showed 100% mortality. When transgenic leaves were fed to cotton bollworm and beet armyworm 100% mortality was observed, whereas there was no mortality observed in the control, and the entire leaf piece was devoured (9).

With the successful introduction of *cry*2Aa2 into the chloroplast genome, the high-dose strategy should be attainable in other crops. This study shows 100% mortality of both Bt-susceptible and Cry1Ac-resistant and Cry2Aa2-resistant tobacco budworm. This is the first report where neonate insects, highly resistant to Bt, were killed using Bt transgenic leaf material even though tobacco budworm is less sensitive to Cry2Aa2 than Cry1Ac. These results are promising when related to reports showing marginal to high levels of cross-resistance to Cry2Aa2 (45,46). This study also shows 100% mortality of cotton bollworm that contrasts with Bt cotton (Cry1Ac) efficacy against cotton bollworm. The inefficient control of cotton bollworm might also result in faster development of Bt resistance because a moderate level of suppression (25-50% mortality) can increase the probability of resistance development (44, 48). In this context, plants expressing *cry*2Aa2 through the chloroplast either singly, or as part of a gene-pyramid with other insect proteins (preferably non-Bt proteins with different modes of action) could become an invaluable tool for resistance management.

Engineering pathogen resistance via the chloroplast genome
Since the beginning of civilization, plant diseases have plagued global crop production. Between 1979 and 1980 India lost up to 60% of its' rice crop due to bacterial rice blight. Between 1988 and 1990, there was a 10.1% loss of the global barley crop due to bacterial pathogens, worth $1.9 billion (50). In the United States, there was an estimated 44,600 metric ton reduction of soybean crops due to bacterial pathogens in 1994. Many efforts have been made to combat these devastating pathogens. Plant breeding was introduced to fight plant diseases (49). However, results were limited due to the ability of the bacteria to adapt and find a way around the defense mechanism. Agrochemicals have been used but their application is limited by their toxicity to humans and the environment (49). With the emergence of molecular biology, researchers have been able to elucidate many of the pathways and products in the plant response to phytopathogens.

In general, the plant defense response can be divided into 3 major categories, early defense (fast), local defense (fast/intermediate) and systemic defense (intermediate to slow, 49). During the early stage, the plant cell is stimulated by contact with pathogen-produced elicitors. Bacterial genes such as *hrp* (hypersensitive response and pathogenicity) or *avr* (avirulence) genes stimulate the plant defense mechanism (50). The most prominent early defense response is the HR (hypersensitive response), which leads to cellular death reducing further infection by the pathogen. Local defense entails cell wall reinforcement, stimulation of secondary metabolite pathways, synthesis of thionins and pathogenesis-related proteins (49). The final phase is known as SAR (systemic acquired resistance), which protects the uninfected regions of the plant. Genetic engineering has allowed for the enhancement of natural defense genes from plants by cloning and over expression in non-host plants. Cloning of resistance (R) genes has been used to protect rice from bacterial leaf blight (49). Pathogenesis-related (PR) genes have been cloned from barley and have shown to provide resistance to *P. syringae* pv. *tabaci* (49). Anti-fungal peptides produced by various organisms have been cloned and studied. While progress made to date is promising for anti-fungal activity (51), bacteria still maintain the ability to adapt to plant defenses.

Plants are not the only species to have problems with pathogenic bacteria. It is common knowledge that the medical community has been fighting a losing battle against pathogenic bacteria for years. The number of multiple drug resistant strains of bacteria is growing, reducing the available choices of antibiotic that can be used. Research continues on multiple fronts to combat antibiotic resistance (52). Cationic antibacterial peptides from mammals, amphibians and insects have gained more attention over the last decade (53). Key features of these cationic peptides are a net positive charge, an affinity for negatively-charged prokaryotic membrane phospholipids over neutral-charged eukaryotic membranes and the ability to form aggregates that disrupt the bacterial membrane or fungal hyphae (54). Given the fact that the outer membrane is an essential and highly conserved part of all bacterial cells, it would seem highly unlikely that bacteria would be able to adapt (as they have against antibiotics) to resist the lytic activity of these peptides.

There are three major peptides with α-helical structures, cecropin from *Hyalophora cecropia* (giant silk moth), magainins from *Xenopus laevis* (African Clawed frog) and defensins from mammalian neutrophils. Magainin and its analogues have been studied as a broad-spectrum topical agent, a systemic antibiotic; a wound-healing stimulant; and an anticancer agent (55). However, the possible agricultural use of magainin-type antimicrobial peptides has not yet been explored. We have recently observed that a synthetic lytic peptide (MSI-99) can be successfully expressed in transgenic chloroplasts (17). The peptide retained its lytic activity against the phytopathogenic bacteria, fungi and multidrug resistant human pathogen, *Pseudomonas aeruginosa*. The anti-microbial peptide (AMP) used in this study was an amphipathic alpha-helical molecule that has an affinity for negatively charged phospholipids commonly found in the outer-membrane of bacteria and fungi. Upon contact with these membranes, individual peptides aggregate to form pores in the membrane, resulting in microbial lysis. Because of the concentration dependent action of the AMP, it was expressed via the chloroplast genome to accomplish high dose delivery at the point of infection. PCR products and Southern blots confirmed chloroplast integration of the foreign genes and homoplasmy. Growth and development of the transgenic plants was unaffected by hyper-expression of the AMP within chloroplasts. *In vitro* assays with T_0, T_1 and T_2 plants confirmed that the AMP was expressed at high levels (21.5 to 43% of the total soluble protein) and retained biological activity against *Pseudomonas syringae,* a major plant pathogen. In addition, leaf extracts from transgenic plants (T_1) inhibited the growth of pre-germinated spores of three fungal species, *Aspergillus flavus, Fusarium moniliforme* and *Verticillium dahliae* by more than 95% compared to untransformed control plant extracts. *In planta* assays with the bacterial pathogen, *Pseudomonas syringae* pv. *tabaci* resulted in areas of necrosis around the point of inoculation in control leaves, while transformed leaves showed no signs of necrosis (200-800 μg of AMP at the site of infection), demonstrating high dose release of the peptide at the site of infection by chloroplast lysis. In planta assays with the fungal pathogen, *Colletotrichum destructivum,* showed necrotic anthracnose lesions in untransformed control leaves, while transformed leaves showed no lesions. T_1 *in vitro* assays against *Pseudomonas aeruginosa* (a multi-drug resistant human pathogen) displayed a 96% inhibition of growth. Hyperexpression may lead to large scale biological production and reduce the cost of chemical synthesis. These results give a new option in the battle against phytopathogenic and drug-resistant human pathogenic bacteria.

Engineering novel pathways via the chloroplast genome

We have recently used the *Bacillus thuringiensis* (Bt) *cry*2Aa2 operon as a model system to demonstrate operon expression and crystal formation via the chloroplast genome (12). *Cry*2Aa2 is the distal gene of a three-gene operon. The *orf* immediately upstream of *cry*2Aa2 codes for a putative chaperonin that facilitates the folding of *cry*2Aa2 (and other proteins) to form proteolytically stable cuboidal crystals (56). Because CRY protein levels decrease in plant tissues late in the growing season or under physiological stress (57), a more stable protein expressed at high levels in the chloroplast throughout the growing season should increase toxicity of Bt transgenic plants to target insects and help eliminate the development of Bt resistance.

Therefore, the *cry*2Aa2 bacterial operon was expressed in tobacco chloroplasts to test the resultant transgenic plants for increased expression and improved persistence of the accumulated insecticidal protein(s). Stable foreign gene integration was confirmed by PCR and Southern blot analysis in T_0 and T_1

transgenic plants. Cry2Aa2 operon derived protein accumulated at 45.3% of the total soluble protein in mature leaves and remained stable even in old bleached leaves (46.1%). This is the highest level of foreign gene expression ever reported in transgenic plants. Exceedingly difficult to control insects (10-day old cotton boll worm, beet army worm) were killed 100% after consuming transgenic leaves. Electron micrographs showed the presence of the insecticidal protein folded into cuboidal crystals similar in shape to Cry2Aa2 crystals observed in *Bacillus thuringiensis*. In contrast to currently marketed transgenic plants with soluble CRY proteins, folded protoxin crystals will be processed only by target insects that have alkaline gut pH; this approach should improve efficacy of Bt transgenic plants. Absence of insecticidal proteins in transgenic pollen eliminates toxicity to non-target insects via pollen. In addition to these environmentally friendly approaches, this observation should serve as a model system for large-scale production of foreign proteins within chloroplasts in a folded configuration enhancing their stability and facilitating single step purification. This is the first demonstration of expression of a bacterial operon in transgenic plants and opens the door to engineer novel pathways in plants in a single transformation event.

Engineering abiotic stress tolerance via the chloroplast genome

Water stress due to drought, salinity or freezing is a major limiting factor in plant growth and development. Trehalose is a non-reducing disaccharide of glucose and its synthesis is mediated by the trehalose-6-phosphate (T6P) synthase and trehalose-6-phosphate phosphatase complex in *Saccharomyces cerevisiae*. In *S. cerevisiae,* this complex consists of at least three subunits performing either T6P synthase (*TPS1*), T6P phosphatase (*TPS2*) or regulatory activities (*TPS3 or TSL1,58,59*) . Trehalose is found in diverse organisms including algae, bacteria, insects, yeast, fungi, animal and plants (60). Because of its accumulation under various stress conditions such as freezing, heat, salt or drought, there is general consensus that trehalose protects against damage imposed by these stresses (61-63). Trehalose is also known to accumulate in anhydrobiotic organisms that survive complete dehydration (64), the resurrection plant (65) and some desiccation tolerant angiosperms (66). Trehalose, even when present in low concentrations, stabilizes proteins and membrane structures under stress (67) because of the glass transition temperature, greater flexibility and chemical stability/inertness.

As pointed out earlier, chloroplast transformation has several advantages over nuclear transformation (1). The difficulty in accomplishing gene containment in nuclear transgenic plants is a serious concern, especially when plants are genetically engineered for drought tolerance, because of the possibility of creating robust drought tolerant weeds and passing on undesired pleiotropic traits to related crops. Chloroplast transformation should also overcome some of the disadvantages of nuclear transformation that result in lower levels of foreign gene expression, such as gene suppression by positional effect or gene silencing (18,19). Therefore, we have recently introduced the yeast *trehalose phosphate synthase (TPS1)* gene into the tobacco chloroplast and nuclear genomes to study resultant phenotypes (28). PCR and Southern blots confirmed stable integration of *TPS1* into the chloroplast genomes of T_1, T_2 and T_3 transgenic plants. Northern blot analysis of transgenic plants showed that the chloroplast transformant expressed 16,966-fold more *TPS1* transcript than the best surviving nuclear transgenic plant. Although both the chloroplast and nuclear transgenic plants showed significant TPS1 enzyme activity, no significant trehalose accumulation was observed in T_0/T_1 nuclear transgenic plants whereas chloroplast transgenic plants showed 15-25 fold higher accumulation of trehalose than the best surviving nuclear transgenic plants. Nuclear transgenic plants (T_0) that showed significant amounts of trehalose accumulation showed stunted phenotype, sterility and other pleiotropic effects whereas chloroplast transgenic plants (T_1, T_2, T_3) showed normal growth and no pleiotropic effects. Chloroplast transgenic plants also showed a high degree of drought tolerance as evidenced by growth of transgenic plants in 6% polyethylene glycol whereas respective control plants were bleached. After 7hr air drying, chloroplast transgenic plants (T_1, T_2, T_3) successfully rehydrated while control plants died. In order to prevent escape of drought tolerance trait to weeds and associated pleiotropic traits to related crops, it is desirable to genetically engineer crop plants for drought tolerance via the chloroplast genome instead of the nuclear genome.

Marker free transgenic plants engineered via the chloroplast genome
        Despite several advantages, one major disadvantage with chloroplast genetic engineering may be the utilization of the antibiotic resistance genes as the selectable marker to confer streptomycin/spectinomycin resistance. When this selection process for chloroplast genetic engineering was first investigated, the mutant 16S rRNA gene that does not bind the antibiotic was used (68). Subsequently, the aadA gene product that inactivates the antibiotic by transferring the adenyl moiety of ATP to spectinomycin /streptomycin was used (69). These antibiotics are commonly used to control bacterial infection in humans and animals. The probability of gene transfer from plants to bacteria living in the gastrointestinal tract or soil may be enhanced by the compatible protein synthetic machinery between chloroplasts and bacteria, in addition to presence of thousands of copies of the antibiotic resistance genes per cell. Also, most antibiotic resistance genes used in genetic engineering originate from bacteria.
        Therefore, betaine aldehyde dehydrogenase (BADH) gene from spinach has been used recently as a selectable marker (70). The selection process involves conversion of toxic betaine aldehyde (BA) by the chloroplast

BADH enzyme to nontoxic glycine betaine, which also serves as an osmoprotectant (71). This enzyme is present only in chloroplasts of a few plant species adapted to dry and saline environments (71, 29). Chloroplast transformation efficiency was 25 fold higher in BA selection than spectinomycin, in addition to rapid regeneration. Transgenic shoots appeared within 12 days in 80% of leaf discs (up to 23 shoots per disc) in BA selection compared to 45 days in 15% of discs (1 or 2 shoots per disc) on spectinomycin selection. Southern blots confirm stable integration of foreign genes into all of the chloroplast genomes (~10,000 copies per cell) resulting in homoplasmy. Transgenic tobacco plants showed 1527-1816% higher BADH activity at different developmental stages than untransformed controls. Transgenic plants were morphologically indistinguishable from untransformed plants and the introduced trait was stably inherited in the subsequent generation. This is the first report of genetic engineering of the chloroplast genome without the use of antibiotic resistance genes. Use of genes that are naturally present in spinach for selection, in addition to gene containment, should ease public concerns or perception of GM crops.

Challenges facing chloroplast genetic engineering
While there are several reports of genetic engineering of the chloroplast genome in tobacco, other major crops (including cereals) have not yet been exploited. One of the major limitations has been the lack of knowledge of chloroplast genome sequences to locate spacer regions and transcriptional units to target site-specific integration of foreign genes. In order to overcome this limitation, Daniell et al. (6) developed the concept of a universal vector that can transform an unknown chloroplast genome because it integrates into a highly conserved region. This concept has been already employed to transform the potato plastid genome using tobacco chloroplast vectors (72).

Another limitation has been the ability to regenerate plants only from embryonic tissues in cereals and not from mesophyll cells. Cells from embryogenic tissues contain only proplastids and not mature plastids. It has been suggested that these plastids are smaller than the size of microprojectiles used for DNA delivery and therefore may pose problems in transformation experiments. Successful expression of chloramphenicol acetyl transferase in proplastids of NT1 cells (73) and β-glucuronidase in proplastids of wheat embryos (74) and other non-green plastids (75) via particle bombardment suggest that particle size may not be a problem in transforming proplastids. Khan and Maliga (75) describe the use of a Fluorescent Antibiotic Resistance Enzyme conferring resistance to spectinomycin/streptomycin (FLARE-S) to detect chloroplast transformation, especially from non-green plastids. FLARE-S was obtained by translational fusion of aminoglycoside adenylyl transferase (aadA) with the green fluorescent protein (GFP) from the jelly fish *Aequorea victoria*. Obtaining stably transformed monocots exhibiting homoplasmy still appears to be a distant accomplishment because only a very small fraction of chloroplasts expressed FLARE-S, when an attempt was made to transform the rice chloroplast genome. Some of the challenges in transforming agronomically useful crops include optimization of tissue culture techniques and the selection process to obtain transgenic plants via particle bombardment, especially from non-green tissues. Even if homoplasmy is not obtained in the first generation, it could be accomplished in subsequent generations by germination of T1 seeds under appropriate selection. In this context, it should be noted that the use of chloroplast integration vectors with ori sequences has been shown to accomplish homoplasmy even in the first round of selection (14). Also, heteroplasmy of chloroplast genomes has been observed in nature (76) and accomplishing homoplasy for the introduced trait may not be always necessary.

Yet another concern is the possibility of yield drag in transgenic crops because of the hyper-expression of foreign genes via the chloroplast genome. High levels of expression of several foreign proteins (up to 50% of total soluble protein) in transgenic chloroplasts have not affected growth rates, photosynthesis, chlorophyll content, flowering, or seed setting (12). Chloroplasts are used to handling such abundant proteins without deleterious effect on productivity. For example, the Calvin cycle enzyme, ribulose bis-phosphate carboxylase/oxygenase (RubisCO) is synthesized as much as 50% of the total soluble protein; such high levels of synthesis have not affected the productivity of crop plants. Indeed, excess RubisCO is constantly made and degraded in chloroplasts. However, long term tests using agronomically important crops grown under field conditions are needed to confirm this observation. Recent success in accomplishing potato plastid transformation should pave the way for studies on such agronomic traits (72). All of these findings augur well for chloroplast genetic engineering of economically useful crops. Thus, several environmentally friendly approaches have been opened for new advances in plant biotechnology and genetic engineering.

References
1. Bogorad, L. *Trends in Biotech.* **2000**,18, 257-263.
2. Daniell, H. *In Vitro Cellular and Developmental Biology-Plant.* **1999**, 35, 361-368.
3. Keeler, K.H.; Turner, C.E.; Bolick, M.R. Duke (Ed). CRC Press, 1996, pp 303-330.
4. Hoyle, B. *J. Nat. Biotech.* **1999**, 17, 747-748.
5. Fox, J.L. *J. Nat. Biotech.* **2000**, 18, 375.
6. Daniell, H.; Datta, R.; Varma, S.; Gray, S.; Lee, S.B. *J. Nat. Biotech.* **1998**, 16, 345-348.
7. Scott, S.E.; Wilkinson, M. *J. Nat. Biotech.* **1999**, 17, 390-392.

8. Daniell, H. Biotechnology and Genetic Engineering Reviews **2000**, 17, 327-352.
9. Kota, M.; Daniell, H.; Varma, S.; Garczynski, F.; Gould, F.; Moar, W. *J. Proc. Natl. Acad. Sci. USA.* **1999**, 96, 1840-1845.
10. Losey, J.E.; Rayor, L.S.; Carter, M.C. *Nature* **1999**, 399, 214.
11. Hodgson, *J. Nat. Biotech.* **1999**, 17, 627.
12. DeCosa, B.; Moar, W.; Lee, S.B.; Miller, M.; Daniell, H. *J. Nat. Biotech.* **2001**, 19, 71-74.
13. Daniell, H. *J. Nat. Biotech.* **1999**, 17, 855-856.
14. Guda, C.; Lee, S.B.; Daniell, H. *Plant Cell Rep.* **2000**,19, 257-262.
15. Staub, J. et al., *J. Nat. Biotech.* **2000**, 18, 333-338.
16. Daniell, H.; Streatfield, S; Wycoff, K. Trends in Plant Science **2001**, 6, 219-226.
17. GeDray, G.; Rajasekaran, K.; Smith, F.; Sanford, J.; Daniell, H. **2001**, Plant Physiology, in press
18. Vaucheret, H. et al. *Plant J.* **1998**, 16, 651-659.
19. De Neve, M. et al. *Mol. Gen. Genetics.* **1999**, 260, 582-592.
20. Navrath, C.; Poirier, Y.; Somerville, C. *Proc. Natl. Acad. Sci.* **1994**, 91, 12760-12764.
21. Ma, J. et al. *Science.* **1995**, 268, 716-719.
22. Ye, X. et al. *Science.* **2000**, 287, 303-305.
23. Holmstrom, K.O.; Mantyla, M.; Wekin, B.; Mandal, A.; Palva, E.T.;Tunnela, O.E.; Londesborough, J. *Nature.* **1996**, 379, 683-684.
24. Goddijn, O.J.M.; Verwoerd, T.C.; Voogd, E.; Krutwagen, W.H.H.; de Graff, P.T.H.M.; Poels, J.; van Dun, K.; Ponstein, A.S.; Damm, B.; Pen, K. *Plant Physiol.* **1997**, 113, 181-190.
25. Romero, C.; Belles, J.M.; Vaya, J.L.; Serrano, R.; Culianz-Macia, F.A. *Planta.* **1997**, 201, 293-297.
26. During, K.; Hippe, S.; Kreuzaler, F.; Schell, J. *Plant Molecular Biology.* **1999**, 15, 281-293.
27. Daniell, H. & Guda, C. *Chemistry and industry.* **1997**, 14, 555-560.
28. Lee, S.B.; Kwon, H.B.; Kwon, S.J.; Park, S.C; Jeong, M.J.; Han, S.E.; Byun,M.O.; Daniell, H. **2001**, in review
29. Nuccio, M.L.; Rhodes, D.; McNeil, S.D.; Hanson, A.D. *Curr. Opinion in Plant Biology.* **1999**, 2,128-134.
30. Puchta, H. *Trends in Plant Science.* **2000**, 5, 273-274.
31. Daniell, H. *Trends in Plant Science.* **2001**, 6, 237-239.
32. Peerenboom, E. *J. Nat. Biotech.* **2000**, 18, 374.
33. Shah, D.M.; Horch, R.B.; Klee, H.J.; Kishore, G.M.; Winter, J.A.; Tumer, E.N.; Hironaka, C.M.; Sanders, P.R.; Gasser, C.S.; Aykent, S.; Siegel, N.R.; Rogers, S.G.; Fraley, R.T. *Science.* **1986**, 233, 478-481.
34. Cioppa, G.D.; Baner, S.C.; Tayler, M.L.; Roshester, D.E.; Klein, B.K.; Shah, D. M.; Fraley, R.T.; Kishore, G.M. *Bio/Technology.* **1997**, 5, 579-584.

35. Gressel, J. *TIBTECH.* **1999**, 17, 361-366.
36. Mariani, C.; DeBeuckeleer, M.; Trueltner, J; Leemans, J.; Goldberg, R.B. *Nature.* **1990**, 347, 737-741.
37. Daniell, H.; Varma, S. *J. Nature Biotechnology* **1998**, 16, 602.
38. Maier, R.M.; Neckermann, K.; Igloi, G.L.; Kössel, H. *J. Mol. Biol.* **1995**, 251, 614-628.
39. Daniell, H. *Methods in Enzymology.* **1993**, 217, 536-556.
40. Daniell, H. *Meth. Mol. Biol.* **1997**, 62: 453-488.
41. Tabashnik, B.E.; Cushing, N.L.; Finson, N.; Johnson, M.W. *J. Econ. Entomol.* **1990**, 83, 1671-1676.
42. Koziel, M.G.; Beland, G.L.; Bowman, C.; Carozzi, N.B.; Crenshaw, R.; Crossland, L.; Dawson, J.; Desai, N.; Hill, M.; Kadwell, S.; Launis, K.; Lewis, K.; Maddox, D.; McPherson, K.; Megjhi, M.R.; Merlin, E.; Rhodes, R.; Warren, G.W.; Wright, M; Evola, S.V. *Bio/Technol.* **1993**, 11, 194-200.
43. Perlak, F.J.; Deaton, R.W.; Armstrong, T.A.; Fuchs, R.L.; Sims, S.R.; Greenplate, J.T.; Fischhoff, D.A. *Bio/Technol.* **1990**, 8, 939-943.
44. Gould, F. *Annu. Rev. Entomol.* **1998**, 43, 701-726.
45. Gould, F.; Martinez-Ramirez, A.; Ferre, J.; Silva, F.J.; Moar, W. *Proc. Natl. Acad. Sci. USA.* **1992**, 89, 7986-7990.
46. Gould, F.; Anderson, A.; Reynolds, A.; Bumgarner, L.; Moar, W. *J. Econ Entomol.* **1995**, 88,1545-1559.
47. Liu, Y.B.; Tabashnik, B.E.; Dennehy, T.J.; Patin, A.L.; Barlett, A.C. *Nature.* **1999**, 400, 519.
48. Tabashnik, B.E.; Liu, Y.B.; Finson, N.; Masson, L.; Heckel, D.G. *Proc. Natl. Acad. Sci. USA.* **1997**, 94,1640-1644.
49. Mourgues, F.; Brisset, M.N.; Cheveau, E. *Trends in Biotechnology.* **1998**, 16, 203-210.
50. Baker, B,; Zambryski, P.; Staskawicz, S.; Dinesh-Kumar, P. *Science.* **1997**, 276, 723-726.
51. Cary,J.; Rajasekaran, K.; Jaynes, J.; Cleveland, T. *Plant Science.* **2000**, 154, 171-181.
52. Persidis, A. *J. Nat. Biotech.* **1999**, 17, 1141-1142.
53. Hancock, R.; Lehrer, R. *TIBTECH.* **1998**, 16, 82-88.
54. Biggin, P. ; Sansom, M. *Biophysical Chemistry.* **1999**, 76, 161-183.
55. Jacob, L.; Zasloff, M. *Ciba Foundation Symposium.* **1994**, 186, 197-223.
56. Ge, B.; Bideshi,D.; Moar,W.; Federici,B. *FEMS Microbiol. Lett.* **1998**, 165, 35-41.
57. Greenplate, J. *J. Econ.Entomol.* **1999**, 92, 1377-1383.
58. Thevelein, J.M.; Hohmann, S. *Trends in Bioscience.* **1995**, 20, 3-10.
59. Singer, M.A. & Lindquist, S. *Trends in Biotech.* **1998**, 16, 460-468.
60. Elbein, A.D. *Adv Carbohyd Chem Biochem.* **1974**, 30, 227-256.
61. Mackenzie, K.F.; Singh, K.K.; Brwon, A.D. *J Gen Microbial.* **1988**, 134,1661-1666.

62. De Vigilio, C.; Hottinger, T.; Dominguez, J.; Boller, T.; Wiekman, A. *Eur J Biochem.* **1994**, 219, 179-186.
63. Sharma, S.C. *Fems Microbiology Letters.* **1997**, 152, 11-15.
64. Crowe J.H., Hoekstra F.A. & Crowe L.M. *Annu Rev Physiol.* **1992**, 54, 579-599.
65. Bianchi, G.; Gamba, A.; Limiroli, R. *Physiol Plantarum.* **1993**, 87, 223-226.
66. Drennan, P.M.; Smith, M.T.; Goldsworthy, D.; Van Staden, J. *J. Plant Physiol.* **1993**, 142, 493-496 (1993).
67. Iwahashi, H.; Obuchi, K.; Fujii, S.; Komatsu, Y. *Cell. Mol. Biol.* **1995**, 41, 763-769.
68. Svab, Z.; Hajdukiewicz P.; Maliga P. *Proc. Natl. Acad. Sci. USA.* **1990**, 87, 8526-8530.
69. Svab Z.; Maliga P. *Proc. Natl. Acad. Sci. USA.* **1993**, 90, 913-917.
70. Daniell, H.; Muthukumar, S.; Lee, S.B. **2001**, Current Genetics, 2001, 39, 109-116.
71. Rathinasabapathy, B.; McCue, K.F.; Gage, D.A.; Hanson, A.D. *Planta.* **1994**, 193, 155-162.
72. Sidorov, V.A.; Kasten, D.; Pang, S.G.; Hajdukiewicz, P.T.J.; staub, J.M.; Nehra, N.S. *Plant J.* **1999**, 19, 209-216.
73. Daniell, H.; Vivekananda, J.; Nielsen, B. L.; Ye, G. N.; Tewari, K. K.; Sanford, J. C. *Proc. Natl. Acad. Sci. USA.* **1990**, 87, 88-92.
74. Daniell, H.; Krishnan, M.; McFadden, B. A. *Plant Cell Rep.* **1991**, 9, 615-619.
75. Khan, S.M.; Maliga, P. *J. Nat. Biotech.* **1999**, 17, 910-915.
76. Frey, J. *Nature.* **1999**, 398:115-116.

Chapter 8

DNA Microchip Technology in the Plant Tissue Culture Industry

K. J. Kunert[1], J. Vorster[1], C. Bester[1], and C. A. Cullis[2]

[1]Botany Department, Forestry and Agricultural Biotechnology Institute,
University of Pretoria, Pretoria 0002, South Africa
[2]Department of Biology, Case Western Reserve University,
Cleveland, OH 44106

Any company involved in micropropagation of plants must be able to demonstrate that the plants produced remain true-to-type as an important part of quality assurance. Modern approaches to detect undesired plant off-types in the *in vitro* propagation process might also include the application of the "DNA-microchip" technology using DNA microarrays carrying hybridization targets isolated from undesirable plant variants. Representational Difference Analysis (*RDA*) has been investigated as a technique for the identification and isolation of potential hybridization targets in a pilot study of somatic embryogenesis in date palm. RDA is a subtractive DNA technique allowing a significant fraction of the plant genome (up to 15%) to be compared between closely related plant lines and to isolate DNA differences between two types of plants. Three difference products from tissue culture-derived plants were isolated by RDA. One product has been further characterized for its potential to monitor genetic variation during the micropropagation process.

INTRODUCTION

The plant tissue culture industry, which represents an estimated $15 billion market with 500 million to 1 billion plants annually produced and an annual growth rate of about 15%, has not been greatly benefited from recent advancements in plant molecular biology. In comparison to traditional plant production, micropropagation of plants via *in vitro* techniques currently represents only a small section of total plant production and is executed predominantly by small to medium-seized commercial companies. Typically, these companies have an annual production in the range of several thousands to several millions of commodity-type plants. There is significant evidence, however, that the importance of plant micropropagation as a source for plant production will dramatically increase through the demand for genetically engineered plants with unique characteristics for which *in vitro* propagation techniques have to be applied as a vital intermediate production step.

In the plant tissue culture industry undesirable plant off-types of low quality are causing severe production losses and consequently will affect the attributes of genetically engineered plants produced via *in vitro* techniques (*1, 2*). Off-type production in plant tissue culture can result from stressful processes (*3, 4, 5*). We have recently started a research program evaluating the idea of using DNA microarrays to address this problem by creating a small prototype diagnostic 'DNA microchip' for off-type detection in tissue culture-derived plants. In this chip-based approach for high throughput screening as a future quality assurance procedure in the plant tissue culture industry, we envisage the hybridization of fluorescence-tagged DNA from test plant DNA to microarrays carrying chemically homogenous plant off-type-derived hybridization targets (*6*). These targets will be isolated by the technique of Representational Difference Analysis (RDA) (*7*). The RDA technique, which belongs to the class of DNA subtractive technologies with the basic concept of comparison between two DNAs and removal of all the sequences held in common between the DNAs, allows isolation of unique DNA sequences that differ between the two DNAs. This technique can monitor 15% or more of the of the plant genome of closely related plant lines in a single experiment.

In this pilot study, we report the application of RDA to identify useful targets from tissue culture-derived date palm plants produced via the process of somatic embryogenesis. Evaluation of the RDA technology on date palm somatic embryogenesis has several benefits. Date palm is among the small number of crops where *in vitro* techniques including somatic embryogenesis have completely replaced traditional vegetative propagation practices. Date palm is, like many important cereals, a monocot plant and off-type markers for date palm might also be applicable in other monocot plants. Further, several tissue culture-derived plant off-types have been recently identified via field evaluation (M. Djerbi, personal communication). Somatic embryogenesis is also the process applied in many transgenic approaches and has the prospect of high multiplication rates, the potential for scaling up in fermentation-type liquid culture and direct delivery to the greenhouse or field as artificial seeds (8).

MATERIALS AND METHODS

Execution of RDA

Date palm material was derived from micropropagated plants of the two cultivars (M and B) produced via the process of somatic embryogenesis using 2,4-D for initiation of embryogenic callus production. Total cellular DNA was isolated from date palm leaves (1 g) applying the Nucleon Phytopure Plant DNA extraction kit (Amersham Life Science, UK) according to the manufacturer's instructions.

To carry out RDA, the general outline for the technique previously described (7) was followed. Two micrograms of each of the DNAs (M and B) were digested with 80 units of the enzyme BamHI. The digests were then extracted with phenol/chloroform, precipitated and resuspended at a concentration of 100μg/ml. The BamHI digests were ligated to a pair of adaptor sequences termed JBam12 (5'-GATCCGTTCATG-3') and JBam24 (5'-ACCGACGTCGACTATCCATGAACG-3'). The ligation products were amplified by the polymerase chain reaction (PCR) as outlined by using the primer (JBam 24) to generate the first round amplicons. Tester DNA was prepared by adding a second adaptor pair NBam 12 and NBam24 (5'-GATCCTCCCTCG-3' and 5'-AGGCAACTGTGCTATCCGAGGGAG-3') to the ends of the first round amplicons from which the J adaptors have been removed.

The hybridization reaction was set up using 40μg of driver DNA (cultivar M) and 0.4μg of tester DNA (cultivar B) (100:1 driver/tester ratio) in a final volume of 4μl of hybridization buffer. The DNA was denatured at 100°C for 10 minutes. Sodium chloride (5M) was added to a final concentration of 1M and the reaction incubated at 67°C for 16 hours. The hybridization mix was then amplified using NBam 24. The first round of amplification was for 10 cycles, followed by digestion of the products by mung bean nuclease. The nuclease-treated product was then amplified for an additional 20 cycles. The resulting amplicons, which are called the first difference product, were used in this study.

The final subtraction products were digested with the appropriate restriction enzyme. The BamHI subtracted products were ligated into BamHI digested pBluescript II (Strategene, USA). The ligation products were transformed into XL1Blue-competent cells and 10 plasmid-containing colonies carrying an insert were selected and digested with BamHI to determine the insert size.

Characterization of difference product

One insert with a size of 141 bp was selected and used to design pairs of primers using a standard design program (Expasy, Switzerland). These primers were then used in a PCR reaction using cultivar B and M DNA as template DNA with an annealing temperature of 65°C. The PCR reactions were carried out in 25µl volumes containing 25ng of total genomic DNA, 15ng of primer, 100mM of each dNTP, 10mM Tris-HCl, pH 8.3, 2mM $MgCl_2$ and 0.5 unit Taq polymerase (Takara, Japan). Amplification was performed on a Perkin Elmer GeneAmp PCR system 9600 with the following program: (i) 94°C for 5 minutes x 1 cycle; (ii) 94°C for 1 minute, 60°C for 1 minute, 72°C for 3 minutes x 35 cycles; (iii) 72°C for 5 minutes x 1 cycle, and optional soak period at 4°C. The products were separated on a 1.5% agarose gel, stained with ethidium bromide and visualized under UV light.

DNA sequencing of amplified PCR-products was done by direct sequencing of the PCR products after purification using a PCR purification kit (Roche, USA). Sequencing was performed using Sequenase (Perkin Elmer, USA) according to the manufacturer's instructions on an automated DNA sequencer (Applied Biosystems, USA).

RESULTS

RDA technique

Figure 1 shows the different steps of the RDA technique. When subtractions of BamHI amplicons were done with cultivar M BamHI amplicons as tester and the cultivar B DNA as driver no difference product was found. The subtraction applying cultivar B as the tester, however, did produce three difference products with a size of 141 bp, 147 bp and 156 bp following one round of subtraction with a 100:1 driver/tester ratio (Figure 2).

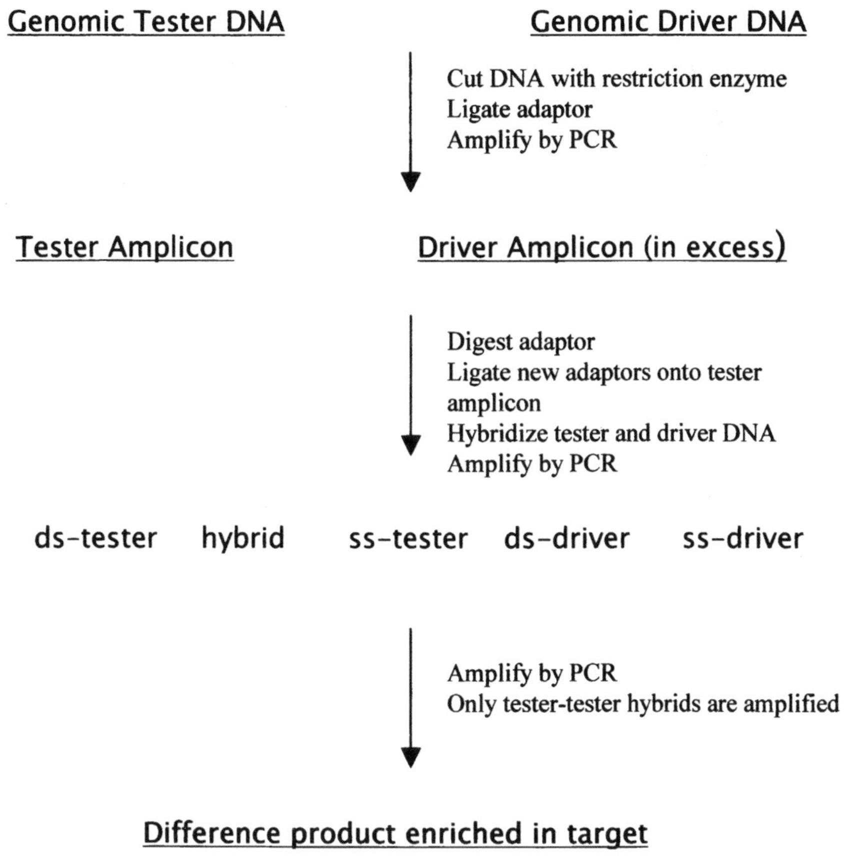

Genomic Tester DNA   Genomic Driver DNA

Cut DNA with restriction enzyme
Ligate adaptor
Amplify by PCR

Tester Amplicon   Driver Amplicon (in excess)

Digest adaptor
Ligate new adaptors onto tester
amplicon
Hybridize tester and driver DNA
Amplify by PCR

ds-tester   hybrid   ss-tester   ds-driver   ss-driver

Amplify by PCR
Only tester-tester hybrids are amplified

Difference product enriched in target

Figure 1. Schematic diagram of RDA

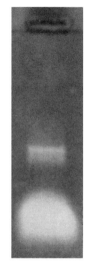

Difference products →

Figure 2. Isolation of cultivar B difference products with a size of 141 bp, 147 bp and 156 bp isolated after first round subtraction using cultivar B DNA as tester and cultivar M DNA as driver.

120 bp →

L M1 M2 M3 M4 M5 M6 B1 B2 B3 B4 B5 B6

Figure 3. Amplification of 120 bp product at 65°C primer annealing temperature in genomic DNAs from different cultivar M (M1-6) and cultivar B (B1-6) plants originating from an in vitro tissue culture process. Lane L represents a 100 bp marker ladder (Roche, Switzerland).

Difference product characterisation

The 141 bp difference product was selected for further characterization. After cloning of the product into the vector pBluescript II, one resulting clone *DP41* was sequenced. Based on the sequence analysis, a primer pair was designed to amplify a 120 bp PCR product. The amplified product was present at a primer annealing temperature of 65°C in all tested plants and did not differentiate between the two cultivars (Figure 3).

Sequence analysis

Alignment of part of the 120 bp amplification product to a number of sequences derived from amplification products of different cultivar B (B1 – B6) and cultivar M DNAs (M1-7) revealed a high degree of homology between the cultivar B and M DNAs (Figure 4).

```
BT  TCTT-GCA-AG-TATCACTGAGGGGGAAGAAGGAGGAGGGGCCTCCGC


B1  TCTT-GCA-AG-TATCACT TAGGGGGAAGGAGGAGGAGGGGCCTCCGC

B2  -CTC-TC-CGCAAATCTTGCAAGTATCAGTAAGAGGAGGGGCCTCCGC
B3  TCTT-GCA-AG-TATCACTGAGGG GAAGGAGGAGGAGGGGCCTCCGC
B4  TCTT-GCACAG-TATCACTGAGGGGGAAGGAGGAGGAGGGGCCTCCGC
B5  TCTT--CA-AG-TATCACTGAGGG-GAAG-AG-AG-AGGG-CCTCCGC
B6  T-TTCGCA-AG-TATCACTGAGGG-GAAGGAGGAGGAGGG-CCTCCGC


M1  TCTT-GCA-AG-TATCACTGAGGGGGAAGGAGGAGGAGGGGCCTAAGC
M2  TCTT-GCA-AG-TATCACTGAGGGGGAAGGAGGAGGAGGGGCCTCTGC
M3  TCTT-GCA-AG-TATCACTGAGGGGGAAGGAGGAGGAGGGGCCTCTGC
M4  TCTT-GCA-AG-TATCACTGAGGGGGAAGGAGGAGGAGGGGCCTCCGC
M5  TCTT-GCA-AG-TATCACTGAGGGGGAAGGAGGAGGAGGGGCCTCCGC
M6  TCTT-GCA-AG-TATCTTTGAGGGGGAAGGAGGAGGAGGGGCCTCCGC
```

Figure 4. DNA sequences derived different cultivar B amplification products (B1-6) and cultivar M amplification products (M1-6) aligned with reference to the cultivar B-derived 141 bp difference product DP41 (BT).

However, several base changes (underlined) were detected in both the cultivar B and the cultivar M DNA sequences. In comparison to all cultivar M sequences, where only a small degree of variation between the sequences was found, a much higher number of base changes consisting of point mutations and deletions were found in cultivar B DNAs. This was most obvious in the cultivar B sequence B2 where almost a 50% change to the sequence of the cultivar B-derived *DP41* difference product was observed. Since cultivar B DNAs derived from randomly selected tissue culture plants propagated from a single mother plant, point mutations and deletions have been introduced via the micropropagation process.

CONCLUSIONS

Will RDA be useful to isolate hybridization targets for a 'DNA microchip?

RDA is a relatively new approach and the technique has not been widely applied for the identification of difference products from genomic plant DNA (*5, 9*). Consequently, information about the usefulness of the technology for the isolation of hybridization targets is still very limited and requires further intensive research efforts. The major advantage of RDA is the specific amplification of fragments exclusively present in one DNA pool allowing an enrichment of tester sequences (*10, 11*). Furthermore, the combination of subtractive hybridization and specific amplification results in DNA difference products of high purity. Like any other technique, RDA is, however, not comprehensive, and identification of differences due to point mutations or very small deletions or insertions in the genome might not be possible. However, as found in our study, several point mutations and base deletions in a sequence, possibly located in a labile DNA region of the plant genome, seems to be sufficient for isolation of differences.

How far are we from the establishment of a DNA microarray?

Figure 5 outlines a typical DNA microarray assay which might be applicable for identification of DNA sequences derived from plant off-types using RDA differences as hybridization targets. However, our knowledge about hybridization targets is currently very limited and we have only isolated and characterized a small number of differences from date palm as reported in this outline and recently also from banana (*9*). We also do not have any indication yet about the degree of variation in tissue culture-derived plants and if identified mutations discovered in cultivar B DNAs have any consequence for the phenotypic appearance or performance of plants. Therefore, we are currently applying RDA on identified date palm off-types and carrying out an extensive DNA sequence analysis program. Furthermore, we have not yet completely evaluated the potential of the RDA technique to identify all possible differences. Since we have only performed RDA with one of the four possible types of amplicons (*7*) and one brand of Taq polymerase, execution with the remaining types of amplicons or different brands of Taq polymerase might allow identification of entirely unique differences. By both isolation of a significant number of differences and by accumulating databases of sequence information as a function of tissue culture processes and off-type characterization, we are confident of identifying DNA sequence patterns involved in *in vitro* plant off-type production, which are useful as hybridization targets.

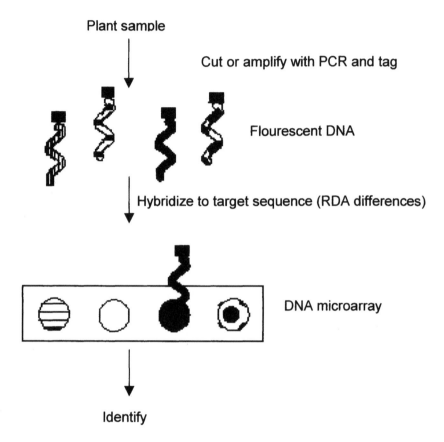

Figure 5. DNA microarray assay for plant off-type detection. DNA from a test plant is either cut by a restriction enzyme or amplified in a PCR reaction and fluorescently labeled. The fluorescence probe is hybridised to a DNA microarray, which might carry RDA-derived specific hybridization targets. Measurement of fluorescence intensity allows the identification of plant-off type DNA.

Would the plant tissue culture industry benefit from a 'DNA microchip'?

The microarray technology is still in its infancy and very expensive, but there is a growing sense that in the future genome analysis will be performed on "microchips". Application of the 'DNA chip' technology specifically in the traditional plant tissue culture industry, represented by small to medium-seized companies will, however, highly depend on whether cost-effective microarray technology can be provided in the future.

DNA microarrays would assist the plant tissue culture industry to optimize its tissue culture processes by monitoring the different *in vitro* propagation steps for plant-off-type production and the early detection of unstable clones in mother plant material entering the tissue culture process. One excellent example is the process of somatic embryogenesis, which is the ability of cultured somatic cells to form embryos required for the *in vitro* mass production of clonal plant material. Plant growth regulators with auxin activity, such as 2,4-D (2,4-dichlorophenoxyacetic acid), which are widely applied in the process of somatic embryogensis, have been implicated in the induction of such variability (*12*) causing chromosome rearrangements, DNA methylation and mutations (*3*). Specifically, in genetically more unstable plants like in the plants of the cultivar B used in our study, variability might be highly prominent. Part of our current research program is therefore directed to investigate in more detail the action of 2,4-D on mutation induction by comparing DNA sequences from embryogenic callus grown on different types and concentrations of auxins. Such variation leads to problems in maintaining regeneration from embryogenic cell lines and in regenerating plants that are morphologically normal and fully fertile reducing dramatically the commercial value of tissue-derived products. This also often means that transgenic plants cannot be used directly in transformation of cereals, which requires long periods of *in vitro* culture on an auxin-containing medium. Microarrays would therefore assist plant biotechnology companies in rapid analysis of such transgenic plants by genome-wide correlations in a single experiment and to show that only the known added gene has changed and not regions responsible for the performance of the plant. This may also ultimately reduce the need for costly field trials.

ACKNOWLEDGMENTS

This work was partially supported by Highveld Biological Ltd, South Africa and NovoMark Technologies LLC, Cleveland, Ohio.

96

References

1. Karp A. Are your plants normal? – Genetic instability in regenerated and transgenic plants" Agro-Food-Industry Hi-Tech **1993**, May/June 7-12.
2. Cassells A.C.; Joyce S.M.; Curry R.F.; McCarthy T.F. Detection of economic variability in micropropagation. In A. Altman, M. Ziv, S. Izhar Eds.; Plant biotechnology and in vitro Biology in the 21st Century. Kluwer Academic Publishers, The Netherlands. 1999; pp. 241-244.
3. Phillips, R.L.; Kaeppler S.M.; Olhoft P. Genetic instability of plant tissue cultures: Breakdown of normal controls. Proceedings of the National. Academy of Science USA **1994**, *91*, 5222-5226.
4. Skirvin, R.M.; McPheeters K.D.; Norton M. Sources and frequency of somaclonal variation. HortScience **1994**, *29*, 1232-1237.
5. Cullis C.; Rademan S.; Kunert K.J. Method for finding genetic markers of somaclonal variation. International publication number WO 99/53100 1999.
6. Lemieux B.; Aharoni A.; Schena M. Overview of DNA chip technology. Molecular Breeding **1998**, *4*, 277-289.
7. Lisitsyn N.; Lisitsyn N.; Wigler M. Cloning the differences between two complex genomes. Science **1993**, *259*, 946-951.
8. Litz R.E.; Gray D.J. Somatic embryogenesis for agricultural improvement. World Journal of Microbiology and Biotechechnology **1995**, *11*, 416-425.
9. Cullis C.; Kunert K.J. Isolation of tissue culture-induced polymorphisms in bananas by representational difference analysis. Acta Horticulturae **2000**, *530*, 421-428.
10. Lisitsyn N.A.; Segre J.A.; Kusumi K. Lisitsyn N.M.; Nadeau J.H.; Frankel W.N.; Wigler M.H.; Lander E.S. Direct isolation of polymorphic markers linked to a trait by genetically directed representational difference analysis. Nature Genetics **1994**, *6*, 57-63.
11. Lisitsyn N.A.; Lisitsyn N.M.; Dalbagni G.; Barker P.; Sanchez C.A.; Gnarra J.; Linehan W.M.; Reid B.J.; Wigler M. Comparative genomic analysis of tumors: Detection of DNA losses and amplification" Proceediings of the National Academy of Sciences USA **1995**, *92*, 151-155.
12. Shoemaker, R.C.; Amberger K.A.; Palmer R.G.; Oglesby L.; Ranch J.P. Effects of 2,4-dichlorophenoxyacetic acid concentration on somatic embryogenesis and heritable variation in soybean (Glycine max L. Mer. R.). In Vitro Cell Development and Biology **1991**, *27*, 84-88.

Chapter 9

Genetic Engineering for Resistance to Phytopathogens

K. Rajasekaran, J. W. Cary, T. J. Jacks, and T. E. Cleveland

Southern Regional Research Center, Agricultural Research Center,
U.S. Department of Agriculture, 1100 Robert E. Lee Boulevard, New
Orleans, LA 70124

Plants are immobile and as such are incapable of escaping
attack by insect and microbial pests. Crop losses due to pests
can be devastating to the point of creating a famine. Fungal
and bacterial pathogens account for the greatest overall losses
associated with plant diseases. Well documented examples
include the infamous Irish potato famine of 1845 due to Potato
Late Blight caused by the fungus *Phytophthora infestans*,
coffee rust caused by the fungus *Hemileia vastatrix* in Ceylon
(Sri Lanka) in the late 1800s and bacterial rice blight disease
caused by *Xanthomonas oryzae* in India in 1979 *(1)*. Not long
ago (in the 1970s) corn production in the USA was threatened
by a highly aggressive new race of *Cochliobolus
heterostrophus* (race T) and wheat production is currently
under threat from Karnal bunt caused by *Telletia indica*. In
addition to yield losses, some fungal pathogens (e.g.
Aspergillus spp., *Fusarium* spp.) cause food and feed safety
concerns because of their ability to produce the potent
mycotoxins, aflatoxin and fumonisin, respectively *(2)*. One of
the primary objectives of conventional plant breeding was to
develop resistance to plant diseases *(3)*. Results, however,
were limited due to the length of time needed to develop
varieties through conventional breeding, the lack of suitable
donor varieties, and the ability of microbes to adapt by

neutralizing plant defense mechanisms. Agrochemicals have been used but their application is limited by several factors including short durability, occurrence of resistance among phytopathogens, toxicity to humans, animals and the fragile environment *(3)*.

Transgenic crops have been developed to combat both insect (e.g. use of Bt gene containing crops) and microbial pathogens. Insect resistant, Bt gene - containing varieties of major crops such as corn, cotton and soybean have made a huge impact in US agriculture and around the world. However, no fungal resistant crops have yet been deregulated for commercial use in spite of numerous field tests *(4)*. Disease resistance has received limited attention for several reasons. The molecular biology of host plant-pathogen interaction is very complex, depending upon single gene or multigenic (quantitative or polygenic) resistance mechanisms and they differ among different races of the same pathogen and different varieties of the same crop species. Quite often, plant defense responses involve the activation of a cascade of multiple, coordinated and apparently complementary responses.

In general, the plant defense response occurs in three gradual phases: early defense (fast), local defense (fast/intermediate) and systemic defense (intermediate to slow) *(3)*. One of the chronologically earliest responses to pathogen invasion is a respiratory burst that produces two reactive oxygen species (ROS) that are microbicidal: superoxide and hydrogen peroxide *(5, 6)*. Bacterial genes such as *hrp* (hypersensitive response and pathogenicity) or *avr* (avirulence) genes stimulate the plant defense mechanism *(7)*. The most prominent early defense response is the hypersensitive response (HR), which leads to localized necrosis thus reducing further spread by the pathogen. Local defense entails cell wall reinforcement, stimulation of secondary metabolite pathways, synthesis of thionins or pathogenesis-related (PR) proteins *(3)*. The final phase is known as systemic acquired resistance (SAR), by which systemic signals are propagated within the plant that protect the uninfected regions of it *(8, 9, 10)*. Researchers are working towards improving defense response of plants by introducing a broad range of genes, from both plant and nonplant sources, to enhance disease resistance during all three phases of defense against fungal and bacterial pathogens (Table 1; also see *(11, 12, 13)*. In the present review, some of the recent advances on developing disease resistant crops using transgenic technology are described. The uses of antimicrobial proteins and peptides to protect the crop plants from phytopathogens are highlighted including parallel work from our laboratory.

Disease resistant genes

Resistance gene (*R*-gene) products may directly or indirectly serve as receptors for pathogen Avr factors *(14)*. The *R*-gene products have been cloned from several plant species and have been successfully transferred to susceptible varieties (Table 1). The examples include *Bs2* from pepper, *Xa21* from rice, *mlo* from barley *(15, 16, 17)*. Transgenic expression of *Xa21* has been shown to confer resistance in susceptible rice varieties to 29 different isolates of the bacterial pathogen, *Xanthomonas oryzae* pv. *oryzae (18)*. Broad-spectrum resistance in crops is thus possible though transgenic expression of *R* genes that are capable of recognizing multiple isolates of a pathogen.

Antifungal proteins

Several studies have demonstrated increased fungal resistance in transgenic plants expressing antifungal proteins (Table 1). These include chitinases, β-1-3-glucanases, ribosome inactivating proteins (RIP), and other pathogenesis-related (PR) proteins (see *3, 19)*. Most filamentous fungal cell walls are made of carbohydrate polymers, chitin and β-1,3-glucans, which are hydrolyzed by the antifungal proteins, chitinase and β-1,3-glucanases in plants. Generally, plants produce these proteins of low-level specificity in response to attack by any microbial pathogen. However, these natural defense mechanisms are not sufficient to ward off microbial invasion and infection. Several laboratories have attempted to boost the defense mechanism by over-expression of the inducible traits or introduce novel factors from other sources. Disease resistance using transgenic strategy often demands a multi-faceted approach. For example, constitutive combined-expression of chitinases and glucanases has resulted in enhanced resistance to phytopathogens in transgenic carrot, tomato and tobacco compared to only one gene *(20, 21, 22)*. A barley RIP was found to be active on fungal ribosomes and thus confer resistance to fungal infections *(23)*. According to Hain et al. *(24)*, the substrates for the enzyme stilbene synthase, p-coumaryl-CoA and malonyl-CoA, are present in most of the plants but not the enzyme itself. Subsequently they transformed tobacco with a grape gene encoding stilbene synthase resulting in synthesis of the phytoalexin, resveratrol, which imparted increased resistance to *Botrytis cinerea*. To augment natural defense mechanisms, Wu et al. *(25)* introduced a glucose oxidase (GO) gene from *A. niger* to generate large amounts of hydrogen peroxide in transgenic potato plants. The transgenic plants showed an increased level of resistance to soft rot caused by *Erwinia carotovora* and to potato late blight caused by *Phytophthora infestans*. Contrary to these observations, a similar attempt with transgenic cottons expressing the *Talaromyces flavus* GO gene resulted in a limited

Table 1. Transgenic plants with antimicrobial activity – selected examples

Gene inserted	Source of gene	Transgenic plants	Phenotype evaluated - Fungal Resistant (FR) or Bacterial Resistant (BR)	References
Antifungal Proteins				
Chitinase (PR-3)	Bean	Tobacco	FR	(61)
	Tobacco	Tobacco	FR	(62)
	Rice	Rice	FR	(63)
	Tomato	Rape	FR	(64)
	Tobacco	Peanut	FR	(65)
	Lycopersicon chilense	Tomato	FR	(66)
	Rice	Cucumber	FR	(67)
	Rice	Grape	FR	(68)
	Trichoderma harzianum	Grape	FR	(69)
		Potato	FR	(70)
	Rice	Rice	FR	(71)
	Rice	Strawberry	FR	(72)
	Rice	Rose	FR	(73)
β-1,3-Glucanase (PR-2)	Soybean	Tobacco	FR	(74)
	Tobacco	Tobacco	FR	(75)
Chitinase + glucanase	Rice	Tobacco	FR	(76)
				(77)
	Barley	Tobacco	FR	(78)
	Tobacco	Tomato	FR	(79)
Agglutinin + chitinase	Urtica dioica, tobacco		FR	(80)
Osmotin	Tobacco	Potato	FR	(81)
Osmotin	Tobacco	Potato	FR	(82)
Germin-like		Wheat	FR	(83)
Hevein	Rubber tree	Indian mustard	FR	(84)
Phytoalexin	Grapevine	Tobacco	FR	(24)
	Grapevine	Rice	FR	(85)

Table I. *Continued*

Gene inserted	Source of gene	Transgenic plants	Phenotype evaluated - Fungal Resistant (FR) or Bacterial Resistant (BR)	References
Phytoalexin	Grapevine	Tobacco	FR	*(86)*
	Grapevine	Alfalfa	FR	*(87)*
thaumatin-like (PR-5)	Tomato	Orange	FR	*(88)*
Antisense construct encoding a rac-related small GTP-binding protein	*Medicago sativa*	Tobacco	FR	*(89)*
Ferredoxin-like (API)	Sweet pepper	Rice	BR	*(90)*
Resistance inducing genes				
R gene *Pto*	Tomato	Tobacco	BR	*(91,92)*
Cf-9	Tomato	Tobacco and potato		*(93)*
Xa21	Rice	Rice	BR	*(18)*
NPR1/NIM1	*Arabidopsis*	*Arabidopsis*	BR/FR	*(94)*
DRR206	Pea	Canola	FR	*(95)*
Glucose oxidase	*Aspergillus niger*	Potato	BR/FR	*(25)*
	A. niger	Cabbage and tobacco	FR/BR	*(96)*
	Talaromyces flavus	Cotton	FR	*(26)*
Chloroperoxidase	*Pseudomonas pyrrocinia*	Tobacco	BR/FR	*(28)*
Puroindolines	Wheat	Rice	FR	*(97)*
Polyphenol oxidase	Tomato	Tomato		*(98)*
Oxalate oxidase	Wheat	Hybrid poplar	FR	*(99)*

Continued on next page.

Table I. *Continued*

Gene inserted	Source of gene	Transgenic plants	Phenotype evaluated - Fungal Resistant (FR) or Bacterial Resistant (BR)	References
Lysozyme	T4 bactriophage	Potato	BR	*(100)*
	human	Tobacco	BR	*(101)*
Lysozyme	Chicken	Potato	BR	*(102)*
Catalase Class II	Tobacco	Potato	FR	*(103)*
Pectate lyase	*Erwinia carotovora*	Potato	BR	*(104)*
Antimicrobial peptides				
Cecropin	Giant Silk Moth	Tobacco	BR	*(36)*
		Tobacco	BR	*(105)*
Cecropin A				*(37)*
Cecropin SB-37		Potato	BR	*(106)*
Cecropin MB-39		Apple	BR	*(107)*
D4E1	cecropin analog - Synthetic	Tobacco	FR/BR	*(108)* *(48)*
Attacin	Cecropia moth	Apple	BR	*(109)* *(110)*
		Potato	BR	*(106)*
Magainin analog	African Clawed Frog	Tobacco		*(111)*
MSI-99	Magainin analog - Synthetic	Tobacco	BR/FR	*(54)*
RIP	Barley seed	Tobacco	FR	*(112)*
		Wheat		*(113)*
Defensin	Radish	Tobacco	FR	*(114)*
	Alfalfa	Potato	FR	*(115)*
Thionin	Barley	Tobacco	BR	*(116)*
	Arabidopsis	*Arabidopsis*	FR	*(117)*
Tachyplesin	Horseshoe crab	Potato	BR	*(118)*

Table I. *Continued*

Gene inserted	Source of gene	Transgenic plants	Phenotype evaluated - Fungal Resistant (FR) or Bacterial Resistant (BR)	References
Cholera toxin	Subunit A1	Tobacco	BR	*(119)*
	Subunit-rotavirus NSP4	Potato		*(120)*
Lactoferrin	Human	Tobacco	BR	*(121)*
Lactoferrin	Human	Potato	BR	*(122)*
AMPs	*Mirabilis jalapa; Amaranthus caudatus*	Tobacco	FR	*(123)*
Sarcotoxin	*Sacrophaga peregrina*	Tobacco	FR	*(124)*
Sarcotoxin 1A		Tobacco		*(125)*
		Tobacco		*(126)*
Cationic peptide chimeras		Potato	FR	*(50)*
		Carrot	FR	*(127)*

antifungal activity against the root pathogen, *Verticillium dahliae (26)*. However, these authors also discovered that the expression of GO in cottons resulted in phytotoxicity and reduced yield. Haloperoxidases (HPOs) such as myeloperoxidase have been shown to convert reactive oxygen species, such as superoxide or hydrogen peroxide, into much more potent ʌtimicrobial agents, hypochlorous acid and peracetic acid, in non-plant systems *(27)* (Figure 1).

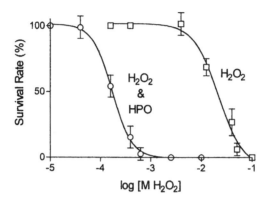

Figure 1. Effect of a metal-requiring haloperoxidase on the lethality of H_2O_2 against Aspergillus flavus spores. Note a 100-fold increase in killing of spores in presence of HPO (T.J. Jacks, unpublished)

Recently, we showed that transgenic tobacco plants and their progenies expressing a novel haloperoxidase (chloroperoxidase) gene from *Pseudomonas pyrrocinia (cpo-p)*, significantly reduced in vitro the number of fungal colonies arising from germinating conidia of *Aspergillus flavus*, and showed greater levels of disease resistance in planta against a fungal pathogen, *Colletotrichum destructivum* that causes anthracnose *(28)* and a bacterial pathogen, *Pseudomonas syringae* pv. *tabaci* (Figure 2, K. Rajasekaran, unpublished).

Transgenic cotton plants expressing the chloroperoxidase gene also demonstrated antifungal activity against *A. flavus*, *V. dahliae* and *Fusarium moniliforme (29)*. Inhibition of *A. flavus* growth by leaf extracts correlated to the amount of CPO-P activity in the leaves (Figure 3).

The enzymically catalyzed reaction due to the presence of nonheme chloroperoxidase in the transgenic plants and the mechanism responsible for imparting resistance to pathogens, however, are currently undetermined *(30, 31)*.

Figure 2. Significant reduction in Fire Blight symptoms caused by the bacterial pathogen Pseudomonas syringae pv. tabaci in the R_1 progeny of transgenic tobacco (var. SR-1) expressing the cpo-p gene (left) as compared to controls – transformed with a minus cpo-p construct (right). From K. Rajasekaran, unpublished.

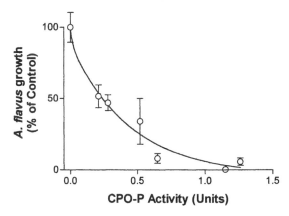

Figure 3. Inhibition of Aspergillus flavus growth by leaf extracts correlated to the amount of CPO-P enzyme activity in the leaves of transgenic tobacco leaves. Each datum represents a separate transgenic plant (n=6). Reproduced from (30)

Chen et al. *(32)* demonstrated a good correlation between high concentrations of a 14-kDa trypsin inhibitor protein present in corn genotypes and resistance to *A. flavus*. They also demonstrated that the trypsin inhibitor protein inhibited the fungal α-amylase thereby reducing the availability of simple sugars for fungal growth. In our laboratory, we have also made a horizontal transfer of the corn trypsin inhibitor gene to tobacco and cotton. Transgenic tobacco and cotton plants expressing the corn trypsin inhibitor protein showed limited antifungal activity in vitro against *V. dahliae* and *A. flavus (33)*

Antimicrobial peptides

Recombinant DNA technologies and plant transformation procedures have been used to introduce and express genes encoding a wide array of antimicrobial peptides (AMPs) in plants in an effort to increase host resistance to plant pathogens (Table 1). Of particular interest has been the identification and characterization of ribosomally synthesized antimicrobial peptides as well as semisynthetic and synthetic peptides. Antimicrobial peptides found in nature are usually less than 50 amino acids in size and are produced by a wide array of organisms *(34)*. These peptides are excellent candidates to augment disease resistance mechanisms in plants due to their i) rapid biocidal or biostatic ability against target cells; ii) activity against a wide spectrum of organisms at low concentrations; and iii) nontoxic nature with respect to mammalian cells *(35, 34)*. In many cases, synthetic analogs of natural antimicrobial peptides or totally synthetic antimicrobial peptides offer even more target specificity, decreased toxicity, increased efficacy at lower concentrations, and reduced degradation by plant proteases than their natural counterparts *(36, 37)*.

Antifungal peptides are produced by numerous organisms including bacteria, fungi, plants, amphibians, and mammals and are linear or disulfide-linked in structure with hydrophobic or amphipathic properties. These peptides have been shown to act either by lysing the fungal cell *(38)*; by aggregating on fungal surfaces and creating pores that cause leakage of ions or other solutes *(39)*; or by interfering with cell wall synthesis *(40)*. AMPs are selective for prokaryotic membranes over eukaryotic membranes due to the predominantly negatively charged phospholipids in the outer leaflet of the prokaryotic membrane *(41, 42, 43)*. Such preference is considered a regulatory function in target selectivity. While the overall charge of the peptide is important, it is known that other features play a role in potency and spectrum of the peptide. The size, sequence, structure (amount of helical content), overall hydrophobicity, amphipathicity and width of the hydrophobic and hydrophilic regions of the peptide have a function in the efficiency of the peptide *(42)*. Due to their diverse modes of action and apparent lack of specific receptors on their target cells, expression of several of these antifungal peptides in transgenic plants should reduce the probability of acquired resistance while simultaneously providing resistance against a number of phytopathogens. Excellent reviews have been published that discuss sources of antimicrobial/antifungal peptides found in nature as well as those produced semisynthetically and synthetically and their targets and modes of action *(42, 44)*. There has been a plethora of reports in the literature pertaining to the in vitro activities of a wide variety of peptides against a number of fungal plant pathogens *(37, 45, 46, 47, 48)*. This section will cover only those studies that have reported on the antifungal activities in planta of peptides that have been genetically engineered into plants using recombinant DNA and plant transformation technologies (Table 1).

Successful transformation of a plant with an antifungal peptide gene and subsequent demonstration of antifungal activity both in vitro and in planta was first reported by Terras et al *(49)*. Tobacco transformed with a binary vector expressing a gene encoding a 5 kDa cysteine-rich protein from radish designated Rs-AFP2 (51 amino acid mature protein) were analyzed for their ability to inhibit growth of the foliar fungal pathogen, *Alternaria longipes*. In vitro assays of crude leaf protein from a transgenic homozygous T2 line expressing Rs-AFP2 demonstrated more than 10-fold higher antifungal activity than extracts of the untransformed control plants. In planta assay of the highest expressing homozygous T2 transgenic line showed an average of 7-8 fold reduction in lesion sizes compared to control plants following inoculation of leaves with *A. longipes*. Both bacterial and fungal phytopathogens were restricted in their ability to infect potato that had been transformed with a synthetic gene encoding a 28 amino acid cecropin-melittin cationic peptide chimera *(50)*. Two cultivars of potato, Russet Burbank and Desiree, were transformed with the synthetic chimera gene, designated msrA1, and transgenic plants expressing the gene were scored for their ability to resist infection by the fungal pathogens, *Phytophthora cactorum* and *F. solani*. By 11 days post infection with *P. cactorum*, control plants were infected from the roots to the tips resulting in the death of the plant while transgenic plants showed no evidence of disease and continued to grow normally. Similar results were obtained from a set of experiments performed with *F. solani*. Interestingly, the expression of msrA1 in cv. Russet Burbank caused morphological changes in the transgenic plants that were not observed in the transgenic Desiree plants or non-transformed controls. These morphological changes were manifested as curly leaves and smaller, branched tubers.

A synthetic, amphipathic, linear peptide (17 amino acids) designated D4E1 has been shown to inhibit the growth, in vitro, of a number of phytopathogenic fungi and bacteria *(48, 51)*; (Table 2). Extracts of leaf tissue from transgenic tobacco expressing the D4E1 gene were shown to significantly reduce the number of fungal colonies arising from germinating conidia of both *A. flavus* and *V. dahliae* *(52)*. Most transgenic plant extracts demonstrated over 90% control of *V. dahliae* compared to control plants. In planta assays of transgenic tobacco expressing D4E1 for resistance to the fungus, *C. destructivum* (causative agent of anthracnose), were also performed. Spores of *C. destructivum* were inoculated onto leaf surfaces of transgenic and control tobacco plants and the severity of disease symptoms was scored after 7 days. Anthracnose severity was

Table 2. Broad-spectrum control of phytopathogens by the synthetic peptide D4E1 *(compiled from 48, 128, 129).*

Alternaria alternata	*Melampsora medusae*
Aspergillus flavus	*Nectria galligena*
A. flavus 70-GFP	*Ophiostoma ulmi*
A. niger	*Penicillium italicum*
A. parasiticus	*Phytophthora cinnamomi*
Cercospora kikuchii	*Phytophthora parasitica*
Claviceps purpurea	*Pseudomonas syringae* pv. *syringae*
Colletotrichum destructivum	*Pythium ultimum* var. *sporangiforium*
Cronartium ribicola	*Rhizoctonia solani*
Fusarium graminearum	*Septoria musiva*
F. moniliforme	*Thielaviopsis basicola*
Fusrium oxysporum	*Verticillium dahliae*
Gremmeniella abietina	*Xanthomonas campestris* pv. *malvacearum* race 18

significantly less on transgenic plants assayed compared to the control plant (Figure 4).

Figure 4. In planta resistance to anthracnose-causing fungus, Colletotrichum destructivum, in transgenic tobacco (cv. Xanthi) plants expressing the antimicrobial, synthetic peptide D4E1. Note the lesions on the leaf of a non-transformed control plant (center) flanked by two transgenic plants (K. Rajasekaran, unpublished).

Preliminary results of in vitro antifungal assays using crude leaf extracts from transformed cotton plants (R_0 and R_1) indicated significant control of *V. dahliae*, a pathogen very sensitive to these two antifungal proteins *(29)*.

In what may be considered the most thorough study of its kind to date, transgenic potato was assayed for resistance to the fungal pathogen, *V. dahliae*, both in the greenhouse and in the field, following transformation with an alfalfa defensin gene, alfALP *(53)*. Plant defensins are family of small (usually 45-54 amino acids) cysteine-rich peptides occurring in various plant species. The alfalfa alfALP gene encodes a 5.6 kDa peptide (45 amino acid mature peptide) that has been shown to inhibit the growth of *V. dahliae* in vitro. Disease resistance in the greenhouse for both control and transgenic potato plants were assayed by dipping their roots in *V. dahliae* spore suspensions before planting in soil. Field plots were artificially infested with *V. dahliae* prior to planting. Disease progression was assessed over a six-week period beginning four weeks after inoculation. In both environments, increased disease resistance to *V. dahliae* was seen in the transgenic plants verses controls. To confirm that enhanced resistance was associated with reduced *V. dahliae* levels in the transgenic plants, stem tissues were sampled and used to determine the number of colony-forming units/ gram dry weight plant tissue. Two potato transgenic lines demonstrating the highest levels of disease resistance also showed a reduction in fungal levels of about six fold compared to control plants. It was concluded from these studies that expression of alfAFP increases field resistance against *V. dahliae* to levels that are equal to, or exceed, those obtained through conventional fumigation of potato. Recently DeGray et al. *(54)* have expressed a gene for a 22mer analog of magainin-2 into the chloroplast genome of tobacco. Crude leaf extracts from transgenic R0, R1 and R2 tobacco plants expressing a 22mer magainin analog (MSI-99) significantly inhibited the growth of pre-germinated spores of three fungal pathogens (*A. flavus*, *F. moniliforme*, and *V. dahliae*). In planta assays demonstrated resistance to a fungal pathogen (*C. destructivum*) and a bacterial pathogen (*P. syringae* pv. *tabaci*). Transformation of chloroplasts is advantageous for several reasons such as increased expression compared to nuclear transformants, prevention of escape of transgenes through pollen, absence of gene silencing and positional effects *(55)*.

Continued identification of novel antimicrobial peptides should provide researchers with the tools necessary to combat a broad-spectrum of phytopathogens including pathogenic fungi that are responsible for enormous pre- and post harvest crop losses annually worldwide. Unfortunately, identification and purification of antimicrobial peptides from living organisms and the subsequent cloning and expression of the genes encoding them is very labor and time intensive. Recent advances in automated peptide synthesis and computer-assisted combinatorial peptide chemistry has made it possible to rapidly formulate, synthesize and screen large numbers of peptides for their ability to inhibit the growth of target microbial pathogens *(56, 57, 42)*. Due to their small size, genes for these synthetic peptides can be readily synthesized and if necessary "stacked" into plant expression vectors with other genes encoding antimicrobial proteins or peptides. In this way, multiple antimicrobial proteins/peptides can be produced within one plant thus reducing the possibility of pathogens acquiring resistance over time. Plants do not produce linear, amphipathic, antimicrobial peptides *(58, 59)*. The availability of synthetic peptides provides a safe and effective compliment to natural antimicrobial proteins and peptides for use in the genetic engineering of plants for resistance to phytopathogens.

In summary, successful introduction of transgenic plants with enhanced pest resistance will not only prevent yield losses but also will be valuable in preventing mycotoxin contamination of food and feed crops *(60)*. Practical applications of these transgenic crops are yet to be realized as an option to combat plant pathogens.

References

1. Nutter, F. W.; Guan, J. Disease Losses, in *Encyclopedia of Plant Pathology*, Maloy, O. C.; Murray, T. D., Eds.; John Wiley & Sons: New York, NY, 2001; pp. 340-351.
2. Sinha.K.K.; Bhatnagar, D.; Eds. *Mycotoxins in Agriculture and Food Safety;* Marcel Dekker, Inc.: New York, NY, 1998.
3. Mourgues, F.; Brisset, M. N.; Chevreau, E. *Trends in Biotechnology* **1998**, *16*, 203-210.
4. http://www.aphis.usda.gov/biotech/petday.html . Feb 4, 2002.
5. Mehdy, M. C. *Plant Physiology* **1994**, *105*, 467-472.
6. Jacks, T.J.; Davidonis, G.H. *Molecular and Cellular Biochemistry* **1996**, *158*, 77-79.
7. Baker, B.; Zambryski, P.; Staskawicz, B.; Dinesh-Kumar, S. P. *Science* **1997**, *276*, 726-733.
8. Ryals, J. A.; Neuenschwander, U. H.; Willits, M. G.; Molina, A.; Steiner, H. Y.; Hunt, M. D. *Plant Cell* **1996**, *8*, 1809-1819.
9. Dempsey, D. A.; Shah, J.; Klessig, D. F. *CRC Critical Reviews in Plant Sciences* **1999**, *18*, 547-575.
10. Kombrink, E.; Schmelzer, E. *European Journal of Plant Pathology* **2001**, *107*, 69-78.
11. Punja, Z. K. *Canadian Journal of Plant Pathology* **2001**, *23*, 216-235.
12. Melchers, L. S.; Stuiver, M. H. *Current Opinion in Plant Biology* **2000**, *3*, 147-152.
13. Rommens, C. M.; Kishore, G. M. *Current Opinion in Biotechnology* **2000**, *11*, 120-125.
14. Staskawicz, B. J. *Plant Physiology* **2001**, *125*, 73-76.
15. Kearney, B.; Staskawicz, B. *Nature* **2002**, *346*, 385-386.
16. Song, W. Y.; Wang, G. L.; Chen, L. L.; Kim, H. S.; Pi, L. Y.; Holsten, T.; Gardner, J.; Wang, B.; Zhai, W. X.; Zhu, L. H.; Fauquet, C.; Ronald, P. *Science* **1995**, *270*, 1804-1805.
17. Wolter, M.; Hollrichter, K.; Salamini, F.; Schultze-Lefert, P. *Molecular and General Genetics* **1993**, *239*, 122-128.
18. Wang, G. L.; Song, W. H.; Ruan, D. L.; Sideris, S.; Ronald, P. C. *Molecular Plant-Microbe Interactions* **1996**, *9*, 850-855.
19. Melchers, L. S.; Stuiver, M. H. *Current Opinion in Plant Biology* **2000**, *3*, 147-152.
20. Jongedijk, E.; Tigelaar, H.; van Roekel, J. S. C.; Bres-Vloemans, S. A.; van den Elzen, P. J. M.; Cornelissen, B. J. C.; Melchers, L. S. *Euphytica* **1995**, *85*, 173-180.

21. van den Elzen, P. J. M.; Jongedijk, E.; Melchers, L. S.; Cornelissen, B. J. C. *Philosophical Transactions of The Royal Society Of London* **1993**, *342*, 271-278.

22. Zhu, Q.; Maher, E. A.; Masoud, S.; Dixon, R. A.; Lamb, C. J. *Bio/Technology* **1994**, *12*, 807-812.

23. Bieri, S.; Potrykus, I.; Futterer, J. *Theoretical and Applied Genetics* **2000**, *100*, 755-763.

24. Hain, R.; Reif, H.-J.; Krause, E.; Langebartels, R.; Kindl, H.; Vornam, B.; Wiese, W.; Schmeltzer, E.; Schreier, P. H.; Stoker, R. H.; Stenzel, K. *Nature* **1993**, *361*, 153-156.

25. Wu, G.; Shortt, B. J.; Lawrence, E. B.; Levine, E. B.; Fitzsimmons, K. C.; Shah, D. M. *Plant Cell* **1995**, *7*, 1357-1368.

26. Murray, F.; Llewellyn, D.; McFadden, H.; Last, D.; Dennis, E. S.; Peacock, W. J. *Molecular Breeding* **1999**, *5*, 219-232.

27. van Pèe, K. H. *Annual Review of Microbiology* **1996**, *50*, 375-399.

28. Rajasekaran, K.; Cary, J. W.; Jacks, T. J.; Stromberg, K. D.; Cleveland, T. E. *Plant Cell Reports* **2000**, *19*, 333-338.

29. Rajasekaran, K.; Cary, J. W.; Jacks, T. J.; Cleveland, T. E., in *Aflatoxin Elimination Workshop*, Robens, J. F., Ed.; USDA, ARS: Beltsville, MD, 2001; pp. 144.

30. Jacks, T. J.; De Lucca, A. J.; Rajasekaran, K.; Stromberg, K.; Van Pee, K. H. *Journal of Agricultural and Food Chemistry* **2000**, *48*, 4561-4564.

31. Jacks, T. J.; Rajasekaran, K.; Stromberg, K. D.; De Lucca, A. J.; van Pèe, K. H. *Journal of Agricultural and Food Chemistry* **2002**, *50*, 706-709.

32. Chen, Z. Y.; Brown, R. L.; Russin, J. S.; Lax, A. R.; Cleveland, T. E. *Phytopathology* **1999**, *89*, 902-907.

33. Rajasekaran, K.; Cary, J. W.; Jacks, T. J.; Chlan, C. A.; Cleveland, T. E., in *Proc. Aflatoxin Elimination Workshop*, Robens, J. F., Ed.; USDA, ARS: Beltsville, MD, 2000; pp. 104.

34. Hancock, R. E. W.; Lehrer, R. *Trends in Biotechnology* **1998**, *16*, 82-88.

35. Jaynes, J. M.; Julian, G. R.; Jeffers, G. W.; White, K. L.; Enright, F. M. *Peptide Research* **1989**, *2*, 157-160.

36. Jaynes, J. M.; Nagpala, P.; Destefano-Beltran, L.; Huang, J. H.; Kim, J.; Denny, T.; Cetiner, S. *Plant Science* **1993**, *89*, 43-53.

37. Cavallarin, L.; Andreu, D.; Segundo, B. S. *Molecular Plant-Microbe Interactions* **1998**, *11*, 218-227.

38. Shai, Y. *Biophysical Journal* **2001**, *80*, 10.

39. Bechinger, B.; Zasloff, M.; Opella, S. J. *Biophysical Journal* **1992**, *62*, 12-14.

40. Debono, M.; Gordee, R. S. *Annual Review of Microbiology* **1994**, *48*, 471-497.
41. Biggin, P.; Sansom, M. *Biophysical Chemistry* **1999**, *76*, 161-183.
42. Tossi, A.; Sandri, L.; Giangaspero, A. *Biopolymers (Peptide Science)* **2000**, *55*, 4-30.
43. Huang, H. W. *Biochemistry* **2000**, *39*, 8347-8352.
44. De Lucca, A. J. *Expert Opinion on Investigational Drugs* **2000**, *9*, 273-299.
45. De Lucca, A. J.; Jacks, T. J.; Broekaert, W. J. *Mycopathologia* **1999**, *144*, 87-91.
46. Lopez-Garcia, B.; Gonzalez-Candelas, L.; Perez-Paya, E.; Marcos, J. F. *Molecular Plant-Microbe Interactions* **2000**, *13*, 837-846.
47. Ali, G. S.; Reddy, A. S. N. *Molecular Plant-Microbe Interactions* **2000**, *13*, 847-859.
48. Rajasekaran, K.; Stromberg, K. D.; Cary, J. W.; Cleveland, T. E. *Journal of Agricultural and Food Chemistry* **2001**, *49*, 2799-2803.
49. Terras, F.-R. G.; Eggermont, K.; Kovaleva, V.; Raikhel, N., V; Osborn, R. W.; Kester, A.; Rees, S. B.; Torrekens, S.; Van Leuven, F.; Vanderleyden, J.; Cammue, B.-P. A.; Broekaert, W. F. *Plant Cell* **1995**, *7*, 573-588.
50. Osusky, M.; Zhou, G.; Osuska, L.; Hancock, R. E.; Kay, W. W.; Misra, S. *Bio/Technology* **2000**, *18*, 1162-1166.
51. De Lucca, A. J.; Bland, J. M.; Grimm, C.; Jacks, T. J.; Cary, J. W.; Jaynes, J. M.; Cleveland, T. E.; Walsh, T. J. *Canadian Journal of Microbiology* **1998**, *44*, 514-520.
52. Cary, J.; Rajasekaran, K.; Jaynes, J. M.; Cleveland, T. E. *Plant Science* **2000**, *154*, 171-181.
53. Gao, A.-G.; Hakimi, S. M.; Mittanck, C. A.; Wu, Y.; Woerner, B. M.; Stark, D. M.; Shah, D. M.; Liang, J.; Rommens, C. M. T. *Bio/Technology* **2000**, *18*, 1307-1310.
54. DeGray, G.; Rajasekaran, K.; Smith, F.; Sanford, J.; Daniell, H. *Plant Physiology* **2001**, *127*, 852-862.
55. Daniell, H.; Khan, M. S.; Allison, L. *Trends in Plant Science* **2002**, *7*, 84-91.
56. Mayo, K. H. *Trends in Biotechnology* **2000**, *18*, 212-217.
57. Blondelle, S. E.; Lohner, K. *Biopolymers (Peptide Science)* **2000**, *55*, 74-87.
58. Broekaert, W. F.; Cammue, B. P. A.; Debolle, M. F. C.; Thevissen, K.; Desamblanx, G. W.; Osborn, R. W. *CRC Critical Reviews in Plant Sciences* **1997**, *16*, 297-323.
59. Garcia-Olmedo, F.; Molina, A.; Alamillo, J. M.; Rodriguez-Palenzuela, P. *Biopolymers* **1998**, *47*, 479-491.

114

60. Dowd, P. F. *Journal of Economic Entomology* **2000,** *93,* 1669-1679.
61. Broglie, K.; Chet, I.; Holliday, M.; Cressman, R.; Biddle, P.; Knowlton, S.; Mauvais, C. J.; Broglie, R. *Science* **1991,** *254,* 1194-1197.
62. Vierheilig, H.; Alt, M.; Neuhaus, J.-M.; Boller, T.; Camejero, V. *Molecular Plant-Microbe Interactions* **1994,** *6,* 261-264.
63. Lin, W.; Anuratha, C. S.; Datta, K.; Potrykus, I.; Muthukrishnan, S.; Datta, S. K. *Biotechnology* **1995,** *13,* 686-691.
64. Grison, R.; Grezes-Besset, B.; Schneider, M.; Lucante, N.; Olsen, L.; Leguay, J. J.; Toppan, A. *Bio/Technology* **1996,** *14,* 643-646.
65. Rohini, V. K.; Rao, K. S. *Plant Science* **2001,** *160,* 889-898.
66. Tabaeizadeh, Z.; Agharbaoui, Z.; Harrak, H.; Poysa, V. *Plant Cell Reports* **1999,** *19,* 197-202.
67. Tabei, Y.; Kitade, S.; Nishizawa, Y.; Kikuchi, N.; Kayano, T.; Hibi, T.; Akutsu, K. *Plant Cell Reports* **1998,** *17,* 159-164.
68. Yamamoto, T.; Iketani, H.; Ieki, H.; Nishizawa, Y.; Notsuka, K.; Hibi, T.; Hayashi, T.; Matsuta, N. *Plant Cell Reports* **2000,** *19,* 639-646.
69. Kikkert, J. R.; Ali, G. S.; Wallace, P. G.; Reisch, B.; Reustle, G. *Acta Horticulturae* **2000,** *528,* 297-303.
70. Lorito, M.; Woo, S. L.; Fernandez, I. G.; Colucci, G.; Harman, G. E.; Pintor-Toro, J. A.; Filippone, E.; Muccifora, S.; Zoina, A.; Tuzun, S.; Scala, F. *Proceedings of the National Academy of Sciences of the United States of America* **1998,** *95,* 7860-7865.
71. Datta, K.; Tu, J.; Oliva, N.; Ona, I.; Velazhahan, R.; Mew, T. W.; Muthukrishnan, S.; Datta, S. K. *Plant Science* **2001,** *160,* 405-414.
72. Asao, H.; Nishizawa, Y.; Arai, S.; Sato, T.; Hirai, M.; Yoshida, K.; Shinmyo, A.; Hibi, T. *Plant Biotechnology* **1997,** *14,* 145-149.
73. Marchant, R.; Davey, M. R.; Lucas, J. A.; Lamb, C. J.; Dixon, R. A.; Power, J. B. *Molecular Breeding* **1998,** *4,* 187-194.
74. Yoshikawa, M.; Tsuda, M.; Takeuchi, Y. *Naturwissenschaften* **1993,** *80,* 417-420.
75. Lusso, M.; Kuc, J. *Physiological and Molecular Plant Pathology* **1996,** *49,* 267-283.
76. Zhu, Q.; Maher, E. A.; Masoud, S.; Dixon, R. A.; Lamb, C. J. *Bio/Technology* **1994,** *12,* 807-812.
77. Jongedijk, E.; Tigelaar, H.; van Roekel, J. S. C.; Bres-Vloemans, S. A.; Dekker, I.; van den Elzen, P. J. M.; Cornelissen, B. J. C.; Melchers, L. S. *Euphytica* **1995,** *85,* 173-180.
78. Jach, G.; Gornhardt, B.; Mundy, J.; Logemann, J.; Pinsdorf, E.; Leah, R.; Schell, J.; Mass, C. *Plant Journal* **1995,** *8,* 97-109.
79. Alexander, D.; Goodman, R.; Gut-Rella, M.; Glascock, C.; Weymann, K.; Friedrich, L.; Maddox, D.; Ahl-Goy, P.; Luntz, T.; Ward, E.; Ryals, J. *Proc Natl Acad Sci* **1993,** *90,* 7327-7331.

80. Does, M. P.; Houterman, P. M.; Dekker, H. L.; Cornelissen, B. J. C. *Plant Physiology* **1999**, *120*, 421-431.
81. Liu, D.; Raghothama, K. G.; Hasegawa, P. M.; Bressan, R. *Proceedings of the National Academy of Sciences of the United States of America* **1994**, *91*, 1888-1892.
82. Zhu, B.; Chen, T. H. H.; Li, P. H.; Zhu, B. L. *Planta* **1996**, *198*, 70-77.
83. Schweizer, P.; Christoffel, A.; Dudler, R. *Plant Journal* **1999**, *20*, 541-552.
84. Kanrar, S.; Venakateshwari, J. C.; Kirti, P. B.; Chopra, V. L. *Plant Science* **2002**, *162*, 441-448.
85. Stark-Lorenzen, P.; Nelke, B.; Hanbler, G.; Muhlbach, H. P.; Thomzik, J. E. *Plant Cell Reports* **1997**, *16*, 668-673.
86. Hunn, B.; Wright, M. S.; Cary, J. W.; Rajasekaran, K.; Johnson, R. M.; Timpte, C. American Society for Microbiology **1999** abstr.
87. Hipskind, J. D.; Paiva, N. L. *Molecular Plant-Microbe Interactions* **2000**, *13*, 551-562.
88. Fagoaga, C.; Rodrigo, I.; Conejero, V.; Hinarejos, C.; Tuset, J. J.; Arnau, J.; Pina, J. A.; Navarro, L.; Pena, L. *Molecular Breeding* **2001**, *7*, 175-185.
89. Schiene, K.; Puhler, A.; Niehaus, K. *Molecular and General Genetics* **2000**, *263*, 761-770.
90. Tang, K. X.; Sun, X. F.; Hu, Q. N.; Wu, A. Z.; Lin, C. H.; Lin, H. J.; Twyman, R. M.; Christou, P.; Feng, T. Y. *Plant Science* **2001**, *160*, 1035-1042.
91. Thilmony, R. L.; Chen, Z.; Bressan, R. A.; Martin, G. B. *Plant Cell* **1995**, *7*, 1529-1536.
92. Rommens, C. M. T.; Salmeron, J. M.; Oldroyd, G. E. D.; Staskawicz, B. J. *Plant Cell* **1995**, *7*, 1537-1544.
93. Hammond-Kosack, K. E.; Tang, S. J.; Harrison, K.; Jones, J. D. G. *Plant Cell* **1998**, *10*, 1251-1266.
94. Cao, H.; Li, X.; Dong, X. N. *Proceedings of the National Academy of Sciences of the United States of America* **1998**, *95*, 6531-6536.
95. Wang, Y.; Fristensky, B. *Molecular Breeding* **2001**, *8*, 263-271.
96. Lee, Y. H.; Yoon, I. S.; Suh, S. C.; Kim, H. I. *Plant Cell Reports* **2002**, *20*, 857-863.
97. Krishnamurthy, K.; Balconi, C.; Sherwood, J. E.; Giroux, M. J. *Molecular Plant-Microbe Interactions* **2001**, *14*, 1255-1260.
98. Li, L.; Steffens, J. C. *Plant Physiology* **1997**, *114*, 1153.
99. Liang, H.; Maynard, C. A.; Allen, R. D.; Powell, W. A. *Plant Molecular Biology* **2001**, *45*, 619-629.
100. Düring, K.; Porsch, P.; Flaudung, M.; Lörz, H. *Plant Journal* **2002**, *3*, 587-598.

101. Nakajima, H.; Muranaka, T.; Ishige, F.; Akutsu, K.; Oeda, K. *Plant Cell Reports* **1997**, *16*, 674-679.
102. Serrano, C.; Arce-Johnson, P.; Torres, H.; Gebauer, M.; Gutierrez, M.; Moreno, M.; Jordana, X.; Venegas, A.; Kalazich, J.; Holuigue, L. *American Journal of Potato Research* **2000**, *77*, 191-199.
103. Yu, D. Q.; Xie, Z. X.; Chen, C. H.; Fan, B. F.; Chen, Z. X. *Plant Molecular Biology* **1999**, *39*, 477-488.
104. Wegener, C.; Bartling, S.; Olsen, O.; Weber, J.; Wettstein, D. v.; Von-Wettstein, D. *Physiological and Molecular Plant Pathology* **1996**, *49*, 359-376.
105. Huang, Y.; Nordeen, R. O.; Di, M.; Owens, L. D.; McBeath, J. H. *Phytopathology* **1997**, *87*, 494-499.
106. Arce, P.; Moreno, M.; Gutierrez, M.; Gebauer, M.; Dell'Orto, P.; Torres, H.; Acuna, I.; Oliger, P.; Venegas, A.; Jordana, X.; Kalazich, J.; Holuigue, L. *American Journal of Potato Research* **1999**, *76*, 169-177.
107. Liu, Q.; Ingersoll, J.; Owens, L.; Salih, S.; Meng, R.; Hammerschlag, F. *Plant Cell Reports* **2001**, *20*, 306-312.
108. Cary, J. W.; Rajasekaran, K.; Jaynes, J. M.; Cleveland, T. E. *Plant Science* **2000**, *154*, 171-181.
109. Reynoird, J. P.; Mourgues, F.; Norelli, J.; Aldwinckle, H. S.; Brisset, M. N.; Chevreau, E. *Plant Science* **1999**, *149*, 23-31.
110. Norelli, J. L.; Aldwinckle, H. S.; Destefano-Beltran, L.; Jaynes, J. M. *Euphytica* **1994**, *77*, 123-128.
111. Li, Q. S.; Lawrence, C. B.; Xing, H. Y.; Babbitt, R. A.; Bass, W. T.; Maiti, I. B.; Everett, N. P. *Planta* **2001**, *212*, 635-639.
112. Logemann, J.; Jach, G.; Tommerup, H.; Mundy, J.; Schell, J. *Bio/Technology* **1992**, *10*, 306-308.
113. Bieri, S.; Potrykus, I.; Futterer, J. *Theoretical and Applied Genetics* **2000**, *100*, 755-763.
114. Terras, F. R. G.; Eggermont, K.; Kovaleva, V.; Raikhel, N. V.; Osborn, R. W.; Kester, A.; Rees, S. B.; Torrekens, S.; Van Leuven, F.; Vanderleyden, J.; Cammue, B. P. A.; Broekaert, W. F. *Plant Cell* **1995**, *7*, 573-588.
115. Gao, A. G.; Hakimi, S. M.; Mittanck, C. A.; Wu, Y.; Woerner, B. M.; Stark, D. M.; Shah, D. M.; Liang, J. H.; Rommens, C. M. T. *Bio/Technology* **2000**, *18*, 1307-1310.
116. Carmona, M. J.; Molina, A.; Fernandez, J. A.; Lopez-FAndo, J. J.; Garcia-Olmedo, F. *Plant Journal* **1993**, *3*, 457-462.
117. Epple, P.; Apel, K.; Bohlmann, H. *Plant Cell* **1997**, *9*, 509-520.

118. Allefs, S. J. H. M.; Jong, E. d.; Florack, D. E. A.; Hoogendoorn, C.; Stiekema, W. J.; De-Jong, E. R. *Molecular Breeding* **1996**, *2*, 97-105.

119. Beffa, R.; Szell, M.; Meuwly, P.; Pay, A.; Vogelilange, R.; Metraux, J. P.; Neuhaus, G.; Meins, F.; Nagy, F. *EMBO Journal* **1995**, *14*, 5753-5761.

120. Arakawa, T.; Yu, J.; Langridge, W. H. R. *Plant Cell Reports* **2001**, *20*, 343-348.

121. Mitra, A.; Zhang, Z. *Plant Physiology* **1994**, *106*, 977-981.

122. Chong, D. K. X.; Langridge, W. H. R. *Transgenic Research* **2000**, *9*, 71-78.

123. De Bolle, M.-F. C.; Osborn, R. W.; Goderis, I. J.; Noe, L.; Acland, D.; Hart, C. A.; Torrekens, S.; Van Leuven, F.; Broekaert, W. F. *Plant Molecular Biology* **1996**, *31*, 993-1008.

124. Mitsuhara, I.; Matsufuru, H.; Ohshima, M.; Kaku, H.; Nakajima, Y.; Murai, N.; Natori, S.; Ohashi, Y. *Molecular Plant-Microbe Interactions* **2000**, *13*, 860-868.

125. Okamoto, M.; Mitsuhara, I.; Ohshima, M.; Natori, S.; Ohashi, Y. *Plant Cell Physiology* **1998**, *39*, 57-63.

126. Ohshima, M.; Mitsuhara, I.; Okamoto, M.; Sawano, S.; Nishiyama, K.; Kaku, H.; Natori, S.; Ohashi, Y. *Journal of Biochemistry* **1999**, *125*, 431-435.

127. Takaichi, M.; Oeda, K. *Plant Science* **2000**, *153*, 135-144.

128. DeLucca, A. J.; Bland, J. M.; Grimm, C.; Jacks, T. J.; Cary, J. W.; Jaynes, J. M.; Cleveland, T. E.; Walsh, T. J. *Canadian Journal of Microbiology* **1998**, *44*, 514-520.

129. Jacobi, V.; Plourde, A.; Charest, P. J.; Hamelin, R. C. *Canadian Journal of Botany* **2000**, *78*, 455-461.

Chapter 10

Engineering Resveratrol Glucoside Accumulation into Alfalfa: Crop Protection and Nutraceutical Applications

N. L. Paiva

Plant Biology Division, The Samuel Roberts Noble Foundation, 2510 Sam Noble Parkway, Ardmore, OK 73401

Stilbenes, including resveratrol (3,5,4'-trihydroxystilbene), are phenolic natural products which accumulate in several plant species, but not in alfalfa (*Medicago sativa*). We have genetically engineered the constitutive accumulation of resveratrol glucoside in transgenic alfalfa leaves and stems. Growth and sporulation of one fungal pathogen was greatly inhibited in transgenic plants, without detrimental effects on plant development. Resveratrol consumption has potential beneficial effects on human health, but there are few dietary sources. Studies in transgenic alfalfa reveal new aspects to be considered in introducing resveratrol synthesis into more human food plants, and provided material for animal tests of the chemopreventive value of such modifications.

Many crop species are infected by bacterial and fungal pathogens, resulting in decreased yields and palatability. In response to pathogen attack, plants often produce low molecular weight antimicrobial compounds called phytoalexins (*1*). In parallel, pathogens have evolved many mechanisms for overcoming these plant defenses (2). Many pathogens have highly specific detoxification enzymes which convert the phytoalexins to nontoxic compounds, some pathogens are no longer sensitive to the host's phytoalexins, while other pathogens suppress or do not activate the host plant's phytoalexin biosynthetic pathways.

One approach to improving the fungal resistance of a crop would be to alter the phytoalexin accumulation, by changing the type of phytoalexin, increasing the level of accumulation, or producing the antifungal compounds prior to infection (3). We have recently altered the phytoalexin profile of alfalfa in just these ways, by introducing a foreign gene encoding the phytoalexin biosynthetic gene, resveratrol synthase. While our initial goal was to improve the fungal resistance of the crop, resveratrol consumption may be beneficial to human and animal health, and our transgenic material may provide useful information for developing novel dietary sources of resveratrol.

Transformation Of Alfalfa With A Resveratrol Synthase Construct

Alfalfa (*Medicago sativa*) is among the top five crops grown in the United States in terms of acres of cultivation, and is worth several billion dollars as a forage crop. Alfalfa (also known as lucerne) is a perennial legume with protein rich leaves (4). The leaves and stems are harvested multiple times during the growing season to produce animal feed rich in high quality protein, vitamins and minerals. Most alfalfa is dried and baled for hay, although some is grazed or fed fresh to dairy cows. An average alfalfa field can produce over 3 tons of hay per acre per year. There is also some use of alfalfa for human food, primarily as alfalfa sprouts for salad greens, and for herbal tea. Through a symbiotic association with *Rhizobium meliloti*, alfalfa can fix up to 200 kg nitrogen per acre per year, thereby providing all of the nitrogen required by the crop and improving the soil for the next crop. Due to the high cost of seed and difficulties associated with planting, alfalfa fields are planted with the expectation the field will remain productive for 3-5 years in order to recover the costs of establishment.

While alfalfa does produce antimicrobial isoflavonoid phytoalexins such as medicarpin and coumestrol following pathogen infection (5,6,7), and higher levels are correlated with improved disease resistance of certain cultivars (8), alfalfa is still plagued by several fungal and bacterial pathogens. Infection can either decrease the quality and value of the hay crop, or kill the plants, decreasing production from a field (9).

Resveratrol (Figure 1), a stilbene phytoalexin, accumulates in a diverse range of plant species, either in response to fungal infection, UV or other stress or constitutively in fruit, bark or roots (10-24). However, resveratrol does not

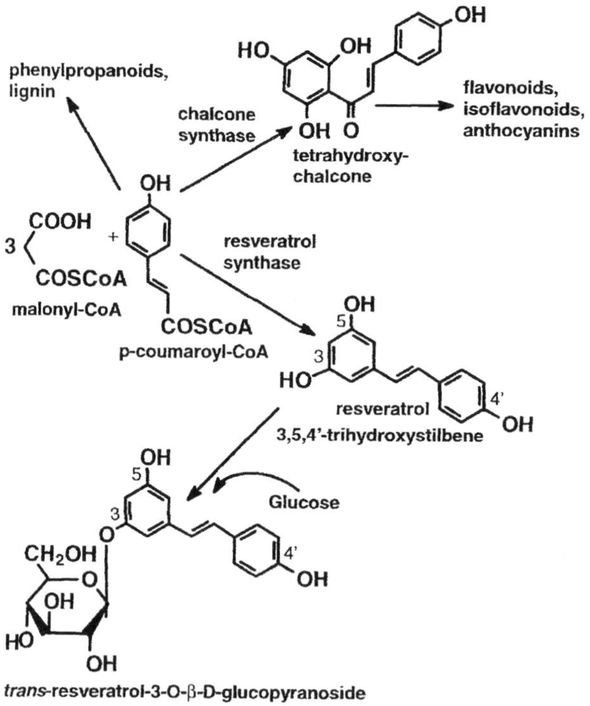

Figure 1: Biosynthetic relationship of the introduced resveratrol and resveratrol glucoside to endogenous flavonoids and phenylpropanoids.

naturally accumulate in alfalfa. Plants which do accumulate resveratrol contain the enzyme resveratrol synthase (also known as stilbene synthase) which condenses three molecules of acetyl coenzyme A with coumaroyl coenzyme A to form a polyketide chain. Rearrangement and decarboxylation yields resveratrol, 3,5,4'-trihydroxystilbene. The identical substrates are used by chalcone synthase, the branchpoint enzyme leading to flavonoids and an enzyme found in all land plants. Chalcone synthase also leads to isoflavonoids in legumes such as alfalfa (*25*). Acetyl coenzyme A and coumaroyl coenzyme A are also precursors of lipids, lignin, and other common plant natural products, and therefore are present in many plant cells at some point in their development. Therefore, the introduction of the single enzyme resveratrol synthase into alfalfa or any land plant could allow the accumulation of resveratrol.

We constructed a binary plant transformation vector containing a stilbene synthase cDNA coding region from peanut (*Arachis hypogaea*) (*26*). Transcription of the gene was under the control of an enhanced cauliflower mosaic virus (CaMV) 35S promoter, an strong constitutive promoter in plants (*27*). The T-DNA portion of the construct was introduced into alfalfa cells via *Agrobacterium tumefaciens*-mediated transformation and transgenic plants were regenerated via somatic embrogenesis (28,29,30). HPLC analysis revealed the presence of a novel peak in acetone extracts of transgenic leaves that was absent from regenerated plants lacking the resveratrol synthase gene (31). While the UV spectrum of the novel peak was highly similar to that of resveratrol, the peak eluted much earlier than an authentic resveratrol standard, suggesting that the resveratrol was conjugated to a more polar molecule. Digestion with crude beta-glucosidase preparations decreased the intensity of the novel peak and released resveratrol. By a combination of HPLC, UV and 1H- and 13C-NMR analyses, the product was identified as trans-resveratrol-3-O-β-D-glucopyranoside (Figure 1) (31); the same glucoside is also known as piceid or polydatin (15,32). The resveratrol glucoside has been isolated from other species as a natural product, but had not previously been identified in a transgenic plant transformed with resveratrol synthase. No free resveratrol was ever observed in the leaf extracts.

The highest concentrations (15-20 ug resveratrol equivalents/g fresh weight) of resveratrol glucoside were observed in young leaves (31). Older leaves contained 1/3 as much. This pattern may be due to the fact that chalcone synthesis is more active in young leaves, which means a good pool of resveratrol precursors must be present, and decreases may be due to turnover of the stilbene product. While the CaMV 35S promoter has been reported to be most active in rapidly dividing cells, the levels of the resveratrol synthase transgene mRNA were high in all leaves, suggesting that transcription of the foreign gene was not the limiting factor in these tissues. In alfalfa stems, the concentration increased from 5 to 7 ug in young stems to 12 to 14 ug resveratrol equivalents/g fresh weight in older stems. A strong increase occurred in the sections of the stem where lignification occurs, again suggesting that coumaroyl CoA was more available in these tissues. Only trace amounts of resveratrol glucoside were detected in alfalfa roots. This may be due to a surprisingly low level of transgene message in the root, as well as very strong competition for precursors by the synthesis of isoflavonoids (25,31).

Increased Pathogen Resistance Of Transgenic Alfalfa Accumulating Resveratrol Glucoside

Phoma medicaginis is an important pathogen of alfalfa, causing "spring black stem" and leaf spot disease in cool wet weather, greatly decreasing yield and quality of alfalfa hay harvested early in the growing season (9). Phoma, in combination with other pathogens, can also slowly kill alfalfa by invading and destroying the crown of the plants. Agar plate bioassays were conducted, wherein the test substance is dissolved in solidified fungal culture medium and the agar inoculated with a piece of fungal mycelium. Equimolar concentrations of resveratrol glucoside purified from alfalfa leaves and free resveratrol (50 ug/ml) inhibited the growth of P. medicaginis equally, approximately 50% (31). Much more significant inhibition was observed following stab-inoculation of the transgenic alfalfa leaves with a suspension of P. medicaginis, even though the leaves contained much lower amounts of resveratrol glucoside than was used in the agar bioassays (*31*). The size of the necrotic lesions in transgenic leaves were reduced 40-80% compared to control alfalfa leaves, and the amount of mycelium (visualized by trypan blue staining) was reduced to 0 to 30% of controls. While numerous pycnidia (the sporulation structure of *P. medicaginis*) formed around the inoculation site in most control leaves, none or very few pycnidia were formed in resveratrol glucoside-accumulating leaves. The average number of pycnidia per inoculated leaf was reduced greater than 98%. Each pycnidia contains multiple spores, which are the main way in which *Phoma* spreads to new leaves and new plants, and are a form in which this pathogen can persist during winter months when the host plant is dormant. Thus, inhibition of pycnidia formation could have a great effect on reducing spread of this pathogen in the field.

Agar plate bioassays demonstrated that several other common alfalfa pathogens were inhibited 35 to 53% by 80 ug resveratrol/ml, including *Fusarium oxysporum* f.sp. *medicaginis, Leptosphaerulina medicaginis, Colletotrichum trifolii, Phytophthora megasperma* var. *medicaginis, Verticillium albo-atrum* (J.D.Hipskind and N.L.Paiva, unpublished results). Testing is underway to determine if resveratrol glucoside-accumulating alfalfa is more resistant than wild-type alfalfa to the stem and leaf pathogens. Due to the low accumulation of product in the roots of transgenic alfalfa, the current expression construct is unsuitable to protect against root pathogens such as *Phytophthora medicaginis*; promoters driving much stronger root expression may allow higher accumulation in the future.

No negative effects on the growth of the resveratrol glucoside-accumulating alfalfa plants was observed; pollination, nodulation by *Rhizobium meliloti,* and flower color were all indistinguishable from that in wild-type lines (*31*). Accumulation of high levels of resveratrol glucoside was also observed in progeny of the primary transgenics, both from self-pollination and crossing of the primary transgenics to more agronomically adapted cultivar (N.L.Paiva, unpublished).

Resveratrol Accumulation In Other Transgenic Plant Species

When our work with alfalfa was initiated, only one report of resveratrol synthase expression in a foreign plant had been published. Hain and co-workers introduced an intact grapevine (*Vitis vinifera*) stilbene synthase gene into tobacco (*Nicotiana tabacum*), such that the foreign gene was still regulated by its original pathogen-inducible grapevine promoter. Transformed plants were much more resistant than control tobacco plants to *Botrytis cinerea* ("grey mold"), providing the first successful example of phytoalexin engineering (*33*). Plants were reported to accumulate as much as 900 ug per g fresh weight 5 days after inoculation with *Botrytis*, with 300-400 ug per g fresh weight being common among transformed lines. The same or very similar pathogen-inducible constructs were later introduced into various species, where reduced lesion sizes were observed following inoculation with appropriate pathogens. Transformed tomato (*Lycopersicon esculentum*) was more resistant to *Phytophthora infestans* (but not *Botrytis cinerea* or *Alternaria solani*) (*34*), transformed rice (*Oryza sativa*) was more resistant to *Pyricularia oryzae* (rice blast pathogen) (*35*), and transformed barley (*Hordeum vulgare*) was more resistant to *Botrytis cinerea* (*36*).

A later study introduced a grapevine stilbene synthase under the control of an enhanced CaMV 35S promoter into tobacco (*37*). This construct is very similar to that used in our alfalfa transformation, and drives constitutive expression of the stilbene synthase in tobacco. The authors reported high levels of free resveratrol in the leaves of transformed tobacco, ranging from 50 ug per gram fresh weight in moderate producers to almost 300 ug per gram fresh weight in high producing lines. They also reported a decrease in pink pigments in flower petals and male sterility due to non-germinating pollen in high producing lines. Since tobacco is known to require flavonols for pollination, and anthocyanins are responsible for the flower color, the authors proposed that the high activity of the stilbene synthase was depleting the cell's reservoir of flavonoid precursors.

In contrast to our results in alfalfa (*31*), wherein only the resveratrol glucoside was observed, only free resveratrol was reported in transgenic tobacco (*37*), despite the similarity of the constructs used. In the above studies, resveratrol in transgenic plants was only assayed using an ELISA assay (*38*) with an antibody raised against resveratrol conjugated to bovine serum albumin (*33,34,37*), or was not measured at all (*35,36*). No direct chemical quantitation of resveratrol or flavonoids was performed, nor was resveratrol glucoside tested for cross-reactivity. Therefore, it is quite possible that these plants were actually accumulating reseveratrol glucoside, which could strongly cross-react with the antibody. Two recent studies support this idea. In our own lab, we have generated transgenic cell cultures of soybean (*Glycine max*) transformed with the same resveratrol synthase vector used in the alfalfa study (*31*). HPLC analysis revealed that these transgenic soybean cultures accumulate low levels of resveratrol glucoside and no resveratrol, while untransformed lines accumulate no resveratrol (J.D. Hipskind and N.L. Paiva, unpublished results). A recently published study with a grapevine stilbene synthase under the control of the CaMV 35S promoter in kiwifuit vines (*Actinidia deliciosa*) also used HPLC to detect the accumulation of only resveratrol glucoside in transformed leaves, with ten times more resveratrol glucoside in young leaves than in older leaves (*39*). We have recently introduced our CaMV 35S: peanut resveratrol synthase vector into tobacco plants and will soon be able to use our HPLC and other phytochemical methods to more closely analyze the resveratrol and flavonoid content in leaves and other plant parts.

Nutraceutical Applications Of Resveratrol Glucoside Accumulation

Many plant phytoalexins and other natural products have been shown to have beneficial activities for use in human medicine. Resveratrol has recently generated a high level of interest among nutritional researchers, due to epidemiological and pharmacological data. Resveratrol has been classified as a "nutraceutical", a non-toxic compound found in food, which although not required like an essential amino acid or used like an enzyme co-factor like many vitamins or minerals, still provides a benefit when consumed as part of a normal diet. The first evidence for the beneficial effects of resveratrol came from studies of the "French paradox", a term based on the observation that although certain groups in France consumed very fatty diets, the rate of cardiovascular disease was unexpectedly low. Researchers credited the consumption of large amounts of red wine in these populations as having a protective effect (18,40). Resveratrol was among the compounds present in red wine which appear to have relevant biological properties, such as acting as a strong antioxidant, inhibiting platelet aggregation, and inhibiting low density lipoprotein (LDL) oxidation (Table I). More recently, resveratrol has been identified in Chinese and South American medicinal roots, some of which were prescribed for treating tumors (17,19). Resveratrol has been shown to have many activities which could help prevent cancer, such as acting as an antioxidant, inhibiting the formation and growth of tumor cells, and inducing apoptosis (programmed cell death) of tumor cells (17; Table I). The estrogenic activity could also contribute to the anticancer properties, as well as aid in maintaining good bone density, as has been suggested for the phytoestrogenic isoflavonoids genistein and daidzein (53,54).

Table I: Reported Biological Activities Of Resveratrol And Resveratrol Glucoside (Piceid)

strong antioxidant (40,41,42)
inhibits platelet aggregation (43)
inhibits LDL oxidation (44)
vasodilator (42,45)
inhibits tumor initiation, promotion and progression (17,18)
inhibits growth of tumor cell lines (17,46)
induces apoptosis in cancer cell lines (47,48)
phytoestrogen (49)
inhibits cyclooxygenase-2 (COX-2) transcription (50)
inhibits IL-2 release (51)
kinase inhibitor (52)
tyrosinase inhibitor (24)

While resveratrol accumulates in many plant species, there are few common dietary sources. In general, constitutive accumulation occurs in roots or bark, where it is thought to serve as a preformed antimicrobial barrier (*19,20,22,23*). Extracts of these tissues have been used as natural medicines but these plants are not routinely consumed as food. Synthesis is induced by fungal infection and UV stress in some species, but this has been most often studied in vegetative tissues and plant cell cultures; such treatments would be impractical or undesirable for inducing synthesis in tissues comsumed by humans.

There are very few examples of resveratrol accumulation in the edible portions of plants. Grapes and grape-derived foods such as wine are the richest and most common source of resveratrol identified to date (Table II). In general the quantity of resveratrol in red wines is higher than that in white wine, due both to the differences in the varieties used and to the fact that the grape skins (high in resveratrol) and stems are left in the initial stages of processing, allowing the resveratrol additional time to elute (*55-58*). However, the amounts of resveratrol and its glucoside in wines and grapes varies with the cultivar, growing environment, and processing and storage conditions (*40,55*). For example, amounts as high as 50 ug/ml have been reported in Portuguese red wines (*56*), but other red wines may contain undetectable levels.

TABLE II. Reported Levels Of Resveratrol And Resveratrol Glucoside (Piceid) In Various Dietary Sources

Grape-derived products:
    Wines (*42,55,56,57*):
        Red: 1-50 ug/ml   (2-10 ug/ml common)
        White: 0-10 ug/ml   (0-2 ug/ml common)
    Grape juice (*16,58*):
        Red: 0.5-5 ug/ml
        White: 0-0.5 ug/ml
Peanut products (*59,60*):
    Fresh: 0.02 (unblemished) – 7.1 ug/g (discolored)
    Roasted: 0.05 ug/g
    Peanut butter: 0.3 ug/g
    Boiled : 5 ug/g
Transgenic resveratrol glucoside-accumulating alfalfa (*31*):
    Fresh: up to 20 ug/g in leaves
    Dry: up to 100 ug/g in mixed leaves and stems

Due to the variations in natural levels, and the high costs or other social and religious barriers to wine consumption, alternative dietary sources of resveratrol could be of interest. Many groups are assaying other foods for resveratrol. Peanuts have been shown to contain very low levels of resveratrol, but the common roasting procedures may destroy resveratrol (Table II) (*59,60*). Also, the presence of resveratrol may be a sign of prior fungal infection or other stresses in peanuts, since resveratrol synthesis is highly induced in this species and damaged peanuts contained higher levels of resveratrol than unblemished peanuts (59).

Many early studies focussed on only the resveratrol aglycone, but recent studies have shown that in wine, the amount of resveratrol present as the glucoside conjugate (piceid) could be higher than the aglycone, especially in grape juice and certain wines (16,56; Table II). The accumulation of the glucoside may protect the aromatic hydroxyl groups from oxidation during harvest and storage, improve uptake in the digestive tract, and be less toxic to the plant.

Alternative sources of resveratrol could be generated by introducing the resveratrol synthase gene into crop species in such a way that resveratrol would accumulate in the edible portion. For example, the construct used for the alfalfa transformation experiments could be introduced into leafy green vegetables such as lettuce or Brassica species, presumably conferring resveratrol glucoside accumulation in tissues that could be consumed fresh or with minimal processing. Other constructs with strong seed promoters could be used to engineer reveratrol accumulation in seeds such as soybeans which already accumulate high levels of isoflavonoids and therefore contain high levels of the needed precursors. The transgenic kiwifruits plants accumulating high levels of resveratrol glucoside in the leaves had not yet set fruit at the time the initial report was published (39); accumulation of resveratrol glucoside in the fruit could produce a highly palatable, novel source of resveratrol.

Transgenic crops such as the resveratrol glucoside-accumulating alfalfa could also serve as a source of resveratrol for extraction for the preparation of dietary supplements. Methods have been developed to express juice from alfalfa for extraction of valuable components, while the solid pulp is still a high quality animal feed. Resveratrol glucoside could be recovered and purified from such juices for human consumption, or the juices could be added to animal feed directly, as is done now to supplement chicken feed with carotenoids (4).

We are currently measuring the resveratrol accumulation and growth of our transformed alfalfa lines in a USDA-approved field test. To date, the lines which were the highest producers in the greenhouse are also the highest producers in the field, and no loss in biomass production is associated with resveratrol accumulation (N.L.Paiva, unpublished data). We also found that the shoots could be dried in standard forage drying ovens, and retain up to 100 ug resveratrol glucoside per gram dry weight of a ground mixture of leaves and stems (Table II). These studies have also allowed us to generate large amounts of pesticide-free alfalfa to use in mouse diet studies. In collaboration with nutritionists at Iowa State University, we are comparing control diets with diets supplemented with either free resveratrol, control alfalfa, and resveratrol-glucoside accumulating alfalfa to see if we can conclusively demonstrate a benefit from engineering resveratrol into edible plants (61). This type of study is easily done with transgenic alfalfa, which produces high amount of biomass and is highly palatable to mice and other lab animals, unlike more easily transformed species such as tobacco or Arabidopsis.

Conclusion

Introduction of a resveratrol synthase coding region from peanut into transgenic alfalfa resulted in the accumulation of high levels of resveratrol glucoside with no detrimental effects on plant development. Unlike previous studies, no free resveratrol was detected in transgenic plant extracts. While agar-plate bioassays indicated that resveratrol was somewhat inhibitory to the growth of *Phoma medicaginis*, plant inoculations demonstrated that accumulation of even small levels of resveratrol glucoside were extremely effective in stopping the development of the pathogen *in planta*. This may be a result of several factors, including: 1) that the antimicrobial compound is accumulated prior to fungal attack unlike most of the natural defenses which are activated following infection, 2) the pathogen has not co-evolved with this foreign phytoalexin, and has no efficiect way to detoxify resveratrol, and 3) the novel phytoalexin is added to the natural defenses of the plant, and acts synergistically with them. In addition to protecting the plant from fungal pathogens, the consumption of resveratrol as a nutraceutical may help reduce the incidence of heart disease and tumors in humans. The same results may be achievable in other crop species, given that the required precursors should be present in all land plants. Unlike many other chemicals used in crop production, the resveratrol glucoside would be classified as "generally regarded as safe" (GRAS) (*62*), since it has been consumed for centuries in the form of wine, grapes and medicinal herbs.

References

1. *Phytoalexins*; J.A. Bailey and J.W. Mansfield Eds.; Halsted Press: New York, NY, 1982, 334 pp.
2. VanEtten, H.D., Matthews, D.E. and Matthews, P.S. *Annu. Rev. Phytopathol.* **1989**, 27, 143-164.
3. Dixon, R.A.; Bhattacharyya, M.K.; Paiva, N.L. In *Advanced Methods in Plant Pathology*; Singh, R.P. and Singh, U.S., Eds.; CRC Press, Boca Raton, FL, 1995; pp 249-270.
4. *Alfalfa and Alfalfa Improvement*; Hanson, A.A.; Barnes, D.K.; Hill, R.R., Eds.; American Society of Agronomy: Madison, WI, 1988; 1084 pp.
5. Higgins, V.J. *Physiol. Plant Pathol.* **1972**, 2, 289-300.
6. Dewick, P.M.; Martin, M. *Phytochemistry* **1979**, 18, 597-602.
7. Paiva, N.L.; Edwards, R.; Sun, Y.; Hrazdina, G.; Dixon, R.A. *Plant Mol. Biol.* **1991**, 17, 653-667.
8. Miller, S.A. In *Plant Cell Culture: A practical approach*. R.A. Dixon, Ed.; IRL Press Ltd.: Washington, D.C., 1985; pp.215-229.

128

9. *Alfalfa Diseases*, 2nd ed.; Stuteville, D.L.; Erwin, D.C., Eds.; American Phytopathological Society: St. Paul, MN, 1990; 104 pp.
10. Bavaresco, L.; Petegolli, D.; Cantu, E.; Fregoni, M.; Chiusa, G.; Trevisan, M. *Vitis* **1997**, 36, 77-83.
11. Schroder, G.; Brown, J.W.S.; Schroder, J. *Eur. J. Biochem.* **1988**, 172, 161-169.
12. Adrian, M.; Jeandet, P.; Bessis, R.; Joubert, J.M. *J. Agric. Food Chem.* **1996**, 44, 1979-1981.
13. Hanawa, F.; Tahara, S.; Mizutani, J. *Phytochemistry* **1992**, 31: 3005-3007.
14. Douillet-Breuil, A.C.; Jeandet, P.; Adrian, M.; Bessis, R. *J. Agric. Food Chem.* **1999**, 47, 4456-4461.
15. Waterhouse, A.L.; Lamuela-Raventós, R.M. *Phytochemistry* **1994**, 37, 571-573.
16. Romero-Perez, A.I.; Ibern-Gomez, M.; Lamuela-Raventos, R.M.; de-la Torre-Boronat, M.C. *J. Agric. Food Chem.* **1999**, 47, 1533-1536.
17. Jang, M.; Cai, L.; Udeani, G.; Slowing, K.; Thomas, C.; Beecher, C.; Fong, H.; Farnsworth, N.; Kinghorn, D.; Mehta, R.; Moon, R.; Pezzuto, J.M. *Science* **1997**, 275: 218-220.
18. Soleas, G.J.; Diamandis, E.P.; Goldberg, D.M. *J. Clin. Lab. Anal.* **1997**, 11, 287-313.
19. Hata, K.; Kozawa, M.; Baba, K. *J. Pharm. Soc. Jap.* **1975**, 95, 211-213.
20. Nonaka, G.; Miwa, N.; Nishioka, I. *Phytochemistry* **1982**, 21, 429-432.
21. Vastano, BC; Rosen, R.T.; Chen, Y.; Zhu, N.Q.; Ho-C.T.; Zhou-Z.Y. *J. Agric. Food Chem.* **2000**, 48, 253-256.
22. Oshima, Y.; Ueno, Y. *Phytochemistry* **1993**, 33, 179-182.
23. Gonzalez-Laredo, R.F.; Chaidez-Gonzalez, J.; Ahmed, A.A.; Karchesy, J.J. *Phytochemistry* **1997**, 46, 175 -176.
24. Likhitwitayawuid, K.; Sritularak, B.; De Eknamkul, W. *Planta Med.* **2000**, 66, 275-277.
25. Dixon, R.A.; Choudhary, A.D.; Dalkin, K.; Edwards, R.; Fahrendorf, T.; Gowri, G.; Harrison, M.J.; Lamb, C.J.; Loake, G.J.; Maxwell, C.A.; Orr, J.D.; Paiva, N.L. In *Phenolic Metabolism in Plants*; Stafford, H.A.; Ibrahim, R.K., Eds.; Plenum Press, New York, NY, 1992; pp 91-138.
26. Tropf, S.; Lanz, T.; Rensing, S.; Schröder, J.; Schröder, G. *J. Mol. Evol.* **1994**, 38, 610-618.
27. Restrepo M.A.; Freed D.D.; Carrington J.C. *Plant Cell* **1990**, 2, 987-998.
28. An, G. *Plant Physiol.* **1986**, 81, 86-91.
29. Bingham, E.T. *Crop Sci.* **1991**, 31, 1098.
30. Bingham, E.T.; Hurley, L.V.; Kaatz, D.M.; Saunders, J.W. *Crop Sci.* **1975**, 15, 719-721.
31. Hipskind, J.D.; Paiva, N.L. *Molecular Plant-Microbe Interact.* **2000**, 13, 551-562.

32. Teguo, P.W.; Decendit, A.; Vercauteren, J.; Deffieux, G.; Mérillon, J-M.. *Phytochemistry* **1996**, 42, 1591-1593.
33. Hain, R.; Reif, H.-J.; Krause, E.; Langebartels, R.; Kindl, H.; Vornam, B.; Wiese, W.; Schmelzer, E.; Schreier, P.; Stöcker, R.; Stenzel, K.. *Nature* **1993**, 361, 153-156.
34. Thomzik, J.; Stenzel, K.; Stöcker, R.; Schreier, P.H.; Hain, R.; Stahl, D. *Physiol. Mol. Plant Path.* **1997**, 51, 265-278.
35. Stark-Lorenzen, P.; Nelke, B.; Hänßler, G.; Mühlbach, H.; Thomzik, J.E. *Plant Cell Rep.* **1997**, 16, 668-673.
36. Leckband, G.; Lörz, H. *Theor. Appl. Genet.* **1998**, 96, 1004-1012.
37. Fischer, R.; Budde, I.; Hain, R. *Plant J.* **1997**, 11, 489-498.
38. Hain, R.; Bieseler, B.; Kindl, H.; Schröder, G.; Stöcker, R. *Plant Mol. Biol.* **1990**, 15, 325-335.
39. Kobayashi, S.; Ding, C.K.; Nakamura, Y.; Nakajima, I.; Matsumoto, R. *Plant Cell Rep.* **2000**, 19, 904-910.
40. *Wine: Nutritional and Therapeutic Benefits*; Watkins, T.R. Ed.; American Chemical Society, Washington, D.C. 1997; 296 p.
41. Tedesco, I.; Russo, M.; Russo, P.; Iacomino, G.; Russo, G.L.; Carraturo, A.; Faruolo, C.; Moio, L.; Palumbo, R. *J. Nutr. Biochem.* **2000**, 11, 114-119.
42. Burns, J.; Gardner, P.T.; O' Neil, J.; Crawford, S.; Morecroft, I.; McPhail, D.B.; Lister, C.; Matthews, D.; MacLean, M.R.; Lean, M.E.J.; Duthie, G.G.; Crozier, A. *J. Agric. Food Chem.* **2000**, 48, 220-230.
43. Orsini, F.; Pelizzoni, F.; Verotta, L.; Aburjai, T.; Rogers, C.B. *J. Nat. Prod.* **1997**, 60, 1082-1087.
44. Pinto, M.C.; Garcia-Barrado, J.A.; Macias, P. *J. Agric. Food Chem.* **1999**, 47: 12, 4842-4846.
45. Fitzpatrick, D.F.; Hirschfield, S.L.; Coffey, R.C. *Amer. J. Physiol.* **1993**, 265, H774-H778.
46. Mgbonyebi, O.P., Russo, J., Russo, I.H. *International J. Oncology* **1998**, 12, 865-869.
47. Clement, M.V.; Hirpara, J.L.; Chawdhury, S.H.; Pervaiz, S. *Blood* **1998**, 92: 996-1002.
48. Thatte, U.; Bagadey, S.; Dahanukar, S. *Cell. Mol. Biol.* **2000**, 46, 199-214.
49. Gehm, B.D.; Mcandrews, J.M.; Chien, P.Y.; Jameson, J.L. *Proc. Natl. Acad. Sci.* **1997**, 94, 14138-14143.
50. Jang, M.; Pezzuto, J.M. *Pharm. Biol.* **1998**, 36 (Suppl.), 28-34.
51. Zhong, M.; Cheng, G.F.; Wang, W.J.; Guo, Y.; Zhu, X.Y.; Zhang, J.T. *Phytomedicine* **1999**, 6, 79-84.
52. Jayatilake, G.S.; Jayasuriya, H.; Lee, E.S.; Koonchanok, N.M.; Geahlen, R.L.; Ashendel, C.L.; Mclaughlin, J.L.; Chang, C.J. *J. Nat. Prod.* **1993**, 56, 1805-1810.

130

53. Herman, C.; Adlercreutz, T.; Goldin, B.R.; Gorbach, S.L.; Hocketrstedt, K.A.V.; Watanabe, S.; Hamalainen, E.K.; Markkanen, M.H.; Mekela, T.H.; Wahala, K.T. *J. Nutr.* **1995**, 125, 757S-770S.
54. Barnes, S. Breast Cancer Research and Treatment **1997**, 46, 169-179.
55. Jeandet, P.; Bessis, R.; Sbaghi, M.; Meunier, P.; Trollat, P. *Am. J. Enol. Vitic.* **1995**, 46, 1-4.
56. Ribeiro de Lima, M.T.; Waffo-Teguo, P.; Teissedre, P.L.; Pujolas, A.; Vercauteren, J.; Cabanis, J.C.; Merillon, J.M. *J. Agric. Food Chem.* **1999**, 47, 2666-2670.
57. Okuda, T.; Yokotsuka, K. *Am. J. Enol. Vitic.* **1996**, 47, 93-99.
58. Ector, B.J.; Magee, J.B.; Hegwood, C.P.; Coign, M.J. *Am. J. Enol. Vitic.* **1996**, 47, 57-62.
59. Sobolev, V.S.; Cole, R.J. *J. Agric. Food Chem.* **1999**, 47, 1435-1439.
60. Sanders, T.H.; McMichael, R.W. Jr.; Hendrix, K.W. *J. Agric. Food Chem.* **2000**, 48, 1243-1246.
61. Au, A.; Stewart, J.W.; Paiva, N.L.; Birt, D.F. *Proc. Annu. Meet. Am. Assoc. Cancer Res.* **2000**, 41: A5371.
62. Maryanski, J.H. In *Genetically Modified Foods: Safety Aspects*. Engel, K.-H.; Takeoka, G.R.; Teranishi, R., Eds.; American Chemical Society, Washington D.C., 1995; p12-22.

Chapter 11

Corn as a Source of Antifungal Genes for Genetic Engineering of Crops for Resistance to Aflatoxin Contamination

Z.-Y. Chen, T. E. Cleveland, R. L. Brown, D Bhatnagar, J. W. Cary, and K. Rajasekaran

[1]Department of Plant Pathology and Crop Physiology, Louisiana State University Agricultural Center, Baton Rouge, LA 70803
[2]Southern Regional Research Center, Agricultural Research Services, U.S. Department of Agriculture, New Orleans, LA 70179

Aflatoxins are toxic, highly carcinogenic secondary metabolites of *Aspergillus flavus* and *A. parasiticus*, produced during fungal infection of a susceptible crop in the field or after harvest, that contaminate food and feed and threaten human and animal health. Natural resistance mechanisms to aflatoxin producing fungi have been identified in corn that could be exploited in plant breeding and/or genetic engineering strategies. Resistant corn lines are being compared to susceptible varieties using proteomics to identify proteins and consequently genes associated with resistance. Antifungal proteins such as ribosomal inactivating proteins, chitinases, protease inhibitors, and lytic peptides have been correlated with increased resistance in corn kernels to invasion by aflatoxigenic fungi. The gene for 14 kD trypsin inhibitor (TI), whose increased levels in corn kernels correlated with enhanced resistance to *A. flavus*, when introduced into tobacco, greatly enhanced resistance in transformed tobacco plants to attack by *Colletotrichum destructivum*. Extracts of cotton embryogenic cultures expressing the TI gene product were shown to cause lysis of germinated *A. flavus* and *Verticillium dahliae* conidia, in vitro. Comparing resistant corn genotypes to susceptible ones through proteomics, may facilitate the identification of several other resistance-associated proteins.

In this chapter we present a brief review of: 1) the history and current status of aflatoxin contamination of corn and strategies being used to eliminate this problem in food and feed; 2) certain antifungal proteins identified in corn and their possible roles in host defense against fungal pathogens; 3) current efforts and progress in identifying proteins associated with host resistance using SDS-PAGE and proteomics approaches; 4) a general understanding of host resistance mechanism(s) in corn against *A. flavus* infection; and finally 5) some progress using antifungal genes identified from corn and other sources in enhancing host plant resistance.

Introduction

Aflatoxins are produced by three species of *Aspergillus*, *A. flavus* Link ex. Fries, *A. parasiticus* Speare, and *A. nomius* Kurtzman, Horn, and Hesseltine (1). Only the first two species appear to be important in the colonization of agricultural commodities. *A. parasiticus* appears to be adapted to a soil environment, being prominent in peanuts, whereas *A. flavus* seems adapted to the foliar environment, being prominent in corn, cottonseed, and tree nuts (2). The dominant aflatoxins produced by *A. flavus* are B_1 and B_2, whereas *A. parasiticus* produces two additional aflatoxins, G_1 and G_2. Aflatoxin B_1 is the most potent naturally occurring carcinogenic substance known (3). Aflatoxins can cause mortality in or reduce the productivity of farm animals (4). Foodstuffs contaminated with aflatoxins also have been associated with increased incidence of liver cancer in humans (5). Therefore, aflatoxin contamination of food and feed not only significantly reduces the value of grains but also poses serious health threats to human and farm animals (6, 7). Food and Drug Administration (FDA) prohibits interstate commerce of feed grains with aflatoxin levels higher than 20 ppb (parts per billion of kernel dry weight).

Infection of corn kernels in the field (preharvest) and during storage (postharvest) by *A. flavus* and subsequent contamination with aflatoxins is a recurrent problem in the southern United States, especially in dry and hot weather conditions. *A. flavus* can grow at very high temperatures (up to 48 °C), and at low water potentials (-35Mpa) (8). These extreme conditions may increase fungal virulence and compromise kernel development, integrity and health, thus, increasing the degree of fungal infection. In the field, *A. flavus* can grow on all kernels especially those damaged by insects or other agents (9), and can infect adjacent intact kernels through the cob tissues (10), the level of aflatoxins in adjacent intact normal-looking kernels can be as high as 4,000 ppb (11). Inside an infected kernel, *A. flavus* usually colonizes the embryo tissue and aleurone layer first, then spreads into the whole endosperm (12).

Measures for controlling aflatoxin contamination in the field focus on prevention of fungal penetration, growth, or subsequent toxin formation in seed. These measures include good cultural practices, harvesting at the optimum stage of maturity and rapid drying after harvesting, and chemical control (13). Cultural practices such as irrigation to reduce plant drought stress can be effective in reducing aflatoxin contamination of corn. However, this practice is not always available or cost-effective for growers, nor does it work all of the time (14). Though applying these management strategies may reduce aflatoxin contamination, the best approach for its elimination is to enhance host-plant resistance through either breeding or genetic engineering (15). This strategy has received much attention recently following discoveries of natural resistance in corn (16).

In the past two decades, several corn genotypes have been identified that show resistance to aflatoxin accumulation in repeated field trials at different locations (10, 17-22). Kernel resistance in some of the genotypes has also been confirmed using a laboratory kernel screening assay (KSA) (12, 23, 24). Unfortunately, these resistant lines have poor agronomical background and the progress of incorporating resistant traits from these lines into commercial corn lines has been slow. Perhaps the greatest hindrance has been the absence of precise physical or chemical factors known to be associated with resistance.

Antifungal Proteins/Genes Identified in Corn

Plants develop a complex variety of defense responses when infected by pathogens. The synthesis of new proteins that can have direct or indirect action on the course of pathogenesis is a common response of plants under pathogen attack. These induced proteins include cell-wall degrading enzymes, proteins with antimicrobial properties, and lytic enzymes. This section will mainly focus on the antifungal proteins and some pathogenesis-related (PR) proteins that have been described in corn.

Chitinases and Beta-1,3-glucanases

The substrate of chitinase, chitin, is a common constituent of fungal cell walls and of the exoskeletons of arthropods, but is unknown in higher plants. Plant chitinases were first isolated from wheat germ and now have been purified from many plants such as barley, beans, and corn (25-27). Some plants may contain multiple chitinases with different molecular masses ranging from 26 to 36 kDa.

The antifungal activity of corn chitinases was first reported by Roberts and Selitrennikoff (27). They reported that the growth of *Trichoderma reesei* and *Phycomyces blakesleeanus* was inhibited by as low as 1 and 3 µg, respectively, of the chitinase isolated from corn (27, 28). In a separate study, two 28 kDa chitinases (CHIT A and CHIT B, class I basic chitinases) isolated from corn were shown to be inhibitory to *T. reesei*, *Alternaria solani*, and *F. oxysporum* (29). Wu et al. (30) reported that the expression level of two chitinase genes, pCh2 and pCh11, was induced by *A. flavus* in aleurone layers and embryos, but not in endosperm tissue. Both pCh2 and pCh11 belong to class I acidic chitinases, and the deduced amino acid sequence of pCh11 resembles a previously reported CHIT D (29). Recently, a 28 kDa chitinase isolated from resistant corn genotype Tex 6 inhibited the growth of *A. flavus* in vitro (31).

Beta-1,3-glucanases have also been found in many plants and range in size from 21-31 kDa (32). This enzyme also releases soluble reducing oligosaccharides from fungal cell walls and has demonstrated antifungal activities against *F. solani* (33). Cordero et al. (34) further observed a coordinated induction of the expression of one beta-1,3-glucanase and three chitinase isoforms in corn seedlings in response to infection by *F. moniliforme*. A recent study by Lozovaya et al. (35) reported that the growth of *A. flavus* was inhibited more by callus of a resistant corn genotype (Tex 6 × Mo17) than by a sensitive genotype (Pa91). This inhibition correlated with the activity levels of beta-1,3-glucanase in the callus and in the culture medium. The presence of the fungus caused an increase in enzyme activity in Tex 6 × Mo17 but not in Pa91 callus. An elevated beta-1,3-glucanase activity in maize kernels was also correlated with lower *A. flavus* infection observed in the resistant genotype (Tex 6) compared with a susceptible one (B73). Further, the antifungal activity of beta-1,3-glucanases was stimulated in the presence of chitinases or other antifungal proteins (33, 36).

Proteinase/Alpha-amylase Inhibitors / Zeamatin

High levels of proteinase inhibitors found in the seeds of many plant species were previously thought to serve as storage or reserve proteins, or as regulators of endogenous enzymes. They were also speculated as defensive agents against attacks by animal predators and insect or microbial pests, since many of the seed protein inhibitors are active in vitro against such microbially produced proteinases (37). Among these inhibitors, some were found to have activity against both trypsin and α-amylase (38, 39). The most extensively studied proteinase inhibitor is trypsin inhibitor (TI). So far, TIs have been isolated from many plants and antifungal activities have been reported for TI proteins from barley (40), corn (41, 42), cabbage (43) and pearl millet (44).

Early in 1973, Halim et al. (41, 45) reported that endosperms of Opaque-2 corn exhibited much higher concentration of TI than normal corn, and growth reduction of *F. moniliforme, A. tenuis,* and *Periconia circinata* of more than 50% was observed when the medium contained 200 to 400 μg/ml of TI isolated from Opaque-2. However, the molecular mass of the isolated TI was not specified. Thus far, three proteinase inhibitors have been isolated from corn. One is the 7 kDa TI protein (46), for which no antifungal activity has been reported. Another is the 22 kDa TI/alpha-amylase inhibitor (39), which showed striking similarity (>50%) to thaumatin II from the fruits of *Thaumatococcus danielli* (47). This protein also shares over 97% homology to a 22 kDa antifungal protein (48) and zeamatin (49)(**Figure 1**). The latter has demonstrated in vitro antifungal properties at low concentrations against *Candida albicans, Neurospora crassa* and *T. reesei.* Zeamatin causes a rapid release of cytoplasmic material from these fungi (49) and also inhibits hyphal growth of *A. flavus* (50). We believe that the 22 kDa antifungal protein, 22 kDa TI/alpha-amylase inhibitor and zeamatin may be encoded by the same gene which varies slightly from one genotype to another or encoded by highly homologous genes of a multigene family.

The third one is the 14 kDa TI protein. It was purified and sequenced in 1984 (51). This TI, which shares no homology to the 22 kDa TI, belongs to the cereal proteinase inhibitor family (37), and was later reported to be an alpha-amylase inhibitor of insects as well (52). Antifungal activities of this TI was first reported by Chen et al. (42). In vitro studies using over-expressed TI purified from *E. coli* found that this recombinant TI inhibited both conidia germination and hyphal growth of nine plant pathogenic fungi studied, including *A. flavus, A. parasiticus,* and *F. moniliforme* (53). This protein is present at high levels in corn genotypes normally resistant to *A. flavus* infection/ aflatoxin contamination, but at low or undetectable levels in susceptible genotypes (42). This same TI also has been reported to be a specific inhibitor of activated Hageman factor (factor XIIa) of the intrinsic blood clotting process (54).

Ribosome-inactivating Proteins (RIPs)

Because of RIPs' ubiquitous distribution and high concentration in plant tissue, it has been speculated to play an important role in defense against parasites (27). RIPs are actually RNA N-glycosidases that catalyze the removal of a specific adenine residue from a conserved 28S rRNA loop required for elongation factor 1 alpha binding. Therefore, RIPs are remarkably potent catalytic inactivators of eukaryotic protein synthesis. All RIPs described to date, including the A-chain of the plant cytotoxin ricin, are polypeptides of 25-32 kDa and share significant amino acid sequence homologies (55). In contrast to previously described RIPs, corn RIP, which is also known as albumin b-32 (56), is synthesized and stored in the kernel as a 34-kDa inactive precursor (pI= 6.5). During germination, this neutral precursor is converted into a basic, active form (pI > 9) by limited proteolysis (55). A two-chain active RIP (comprised of 16.5- and 8.5-kDa fragments that remain

```
                 1                15 16              30 31             45 46               60
1 JS0464         -------------    ------AVFTVNQC    PFTVWAASVPVGGGR    QLNRGESWRITAPAG    39
2 P13867         -------------    ------AVFTVNQC    PFTVWAASVPVGGGR    QLNRGESWRITAPAG    39
3 T02075         MAGSVAIVGIFVALL  AVAGEAAVFTVVNQC   PFTVWAASVPVGGGR    QLNRGESWRITAPAG    60

                 61               75 76             90 91            105 106             120
1 JS0464         TTAARIWARTGCGFD  ASGRGSCRTGDCGGV   QCTGYGRAPNTLAE    YALKQFNNLDFFDIS    99
2 P13867         TTAARIWARTGCGFD  ASGRGSCRTGDCGGV   QCTGYGRAPNTLAE    YALKQFNNLDFFDIS    99
3 T02075         TTAARIWARTGCGFD  ASGRGSCRTGDCGGV   QCTGYGRAPNTLAE    YALKQFNNLDFFDIS   120

                 121             135 136            150 151           165 166            180
1 JS0464         GDGFNVP SFLPDG   GSGCSRGPRCAVDVN   ARCPAELRQDGVCNN   ACPVFKKDEYCCVGS   159
2 P13867         GDGFNVP SFLPDG   GSGCSRGPRCAVDVN   ARCPAELRQDGVCNN   ACPVFKKDEYCCVGS   159
3 T02075         GDGFNVP SFLPDG   GSGCSRGPRCAVDVN   ARCPAELRQDGVCNN   ACPVFKKDEYCCVGS   180

                 181             195 196            210 211           225 226
1 JS0464         AAN CHPTNYSRYFK  GQCPDAYSYPKDDAT   STFTCPAGTNYKVVF   CP   206
2 P13867         AAN CHPTNYSRYFK  GQCPDAYSYPKDDAT   STFTCPAGTNYKVVF   CP   206
3 T02075         AANP CHPTNYSRYFK GQCPDAYSYPKDDAT   STFTCPAGTNYKVVF   CP   227
```

Figure 1. Amino acid sequence homology comparisons of corn 22 kDa antifungal protein (JS0646), 22 kDa zeamatin (T02075) and the 22 kDa alpha-amylase/trypsin inhibitor protein (P13867). The non-conserved amino acid residues were highlighted.

tightly associated) is produced from this processing event. It is interesting to note that in corn the expression of this RIP was recently found to be under the control of transcriptional activator Opaque-2 (57). Purified barley RIPs exhibited antifungal activity in vitro against *Alternaria alternata*, *Phycomyces blakesleeanus* and *T. reesei* (27). In a recent study the RIP purified from corn also inhibited the hyphal development of *A. flavus* (50).

Other pathogenesis related (PR) proteins

PR proteins are a heterogenous group of proteins produced during pathogen invasion. The PR proteins can be grouped into five classes according to their function and sequence similarities (58). PR protein classes 2, 3 and 5 contain chitinases, beta-1,3-glucanases, and thaumatin-like proteins, respectively, all of which have been described above. Class 1 contains related proteins found in a wide variety of plants with molecular masses of approximately 15 kDa that are induced on infection by tobacco mosaic virus. Class 4 includes acidic proteins with MW of 13-14.5 kDa that have sequence similarity with wound-induced proteins of potato. The mechanism of action of classes 1 and 4 PR proteins remains unknown.

After pathogen infection of corn leaves, the PR-1 and PR-5 genes are induced more rapidly and more strongly in an incompatible than in a compatible interaction (59). The predicted size of the mature corn PR-1 and PR-5 polypeptides are 15, and 15.7 kDa, respectively, and both are acidic proteins. The PRms protein is induced by fungal infection in germinating corn seeds, but not in vegetative tissues of the germinating seedlings nor in leaves or roots of the mature plants (60). The functions of these proteins, however, are still unknown.

Current Efforts and Progress in Identifying Potential Resistance Factors (Proteins/Genes)

Potential Protein Markers Identified Using SDS-PAGE

Using one dimensional (1-D) SDS-PAGE to compare kernel proteins of corn genotypes resistant or susceptible to *A. flavus* infection or aflatoxin production, several proteins, whose levels of expression could be associated with resistance to *A. flavus* infection or aflatoxin production, were recently identified. One of these proteins is the 14 kDa trypsin inhibitor (TI) (42) (see above section). The mode of action of this TI on fungal conidia germination and hyphae growth may be partially due to its inhibition on fungal alpha-amylase activity and production (61). The reduction of extracellular fungal α-amylase by TI may limit the availability of simple sugars needed for fungal growth.

The expression of two embryo globulin proteins (GLB1 and GLB2) was also found to be associated with kernel resistance in a recent study (62). Huang et al. (63) compared one resistant with one susceptible genotype and found two proteinaceous fractions inhibitory to aflatoxin formation from resistant genotype Tex-6. However, the ability to identify more resistance associated proteins using 1-D gels is limited due to its low resolution.

Proteomics to Identify Potential Protein Markers

Proteomics is the study of proteins complement expressed by a genome (64). Due to its increased resolution and sensitivity, 2-D polyacrylamide gel electrophoresis is a powerful tool for the analysis and detection of proteins from complex biological sources (65, 66). Recent advances in technologies, such as the use of immobilized pH gradient (IPG) gel strips and the latest sophisticated computerized 2-D gel analysis softwares (67, 68), have significantly increased the reproducibility, reliability and accuracy of 2-D gel electrophoresis. For these reasons, 2-D PAGE has been increasingly used to identify unique and/or developmentally regulated proteins (69, 70).

Recently, endosperm and embryo proteins from several resistant and susceptible genotypes were compared using large format 2-D gel electrophoresis (62, 71). Over 1000 spots were routinely detected in gels containing embryo proteins, and 600 to 800 spots for gels with endosperm proteins. Preliminary comparisons of reproducibly detected spots have found both quantitative and qualitative differences between resistant and susceptible genotypes (**Figure 2**).

Several protein spots, either unique or 5 fold upregulated, were recently isolated from preparative 2-D gels and analyzed using ESI-MS/MS after in-gel digestion with trypsin (71). Amino acid sequence of a few spots showed homology to GLB1 or GLB2 of corn, and some previously reported stress-responsive proteins, such as group 3 late embryogenesis abundant (LEA3) protein, aldose reductase, a glyoxalase I protein, and a 16.9 kDa heat shock protein (71). The aldose reductase appears to be involved in sorbitol biosynthesis and osmoregulation in the embryo (72). There were also a few proteins that showed no significant sequence homology to known sequences in databases. We are in the process of ascertaining the possible roles of these identified proteins in host fungal defense.

Resistant   **Susceptible**

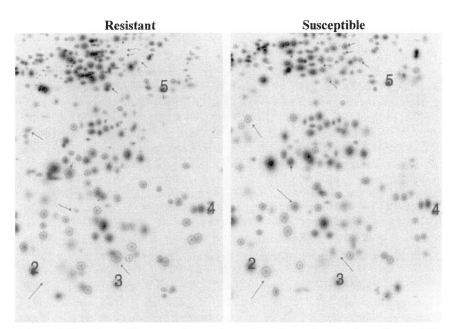

Figure 2. Protein profile comparisons of corn kernel protein using large format 2-D polyacrylamide gels. Shown are partial composite gels generated from resistant and susceptible corn genotypes. Numbers on the gels represent common proteins (anchors) used to align gels for matching. Circles indicate the spots are present in all the gels. Arrows indicate the protein spot differences between resistant and susceptible composite gels.

Understanding of Host Resistance Mechanisms

In the past several years, progress has been made in the understanding of host resistance mechanisms in corn against *A. flavus* infection and aflatoxin production as well as in identifying factors contributing to kernel resistance. These include both physical and biochemical factors strategically distributed both on the kernel surface (pericarp) and inside the kernels.

Pericarp Resistance

Guo *et al.* (23) found that removal of kernel wax and cutin layers increases aflatoxin accumulation in resistant corn population GT-MAS:gk. Further studies by Russin et al (73) found that GT-MAS:gk contains unique wax component. Recently, GT-MAS:gk was compared to 12 susceptible corn genotypes for differences in kernel wax using TLC and GC-MS. Gembeh et al. (74, 75) confirmed that GT-MAS:gk kernel wax contains a unique compound associated with a TLC band that is missing in all susceptible corn genotypes, and further found that GT-MAS:gk lacks another wax band, present in all susceptible genotypes. Using GC-MS, the wax band unique to GT-MAS:gk was determined to contain high levels of ethyl-hexadecanoate and phenolic compounds. It was suggested that these compounds may contribute to inhibition of *A. flavus* growth in GT-MAS:gk.

Subpericarp Resistance-Constitutive Proteins

Many studies have documented the involvement of antifungal proteins in conferring host resistance. We recently found that resistant corn genotypes contain high levels of the 14 kDa TI and a 28 kDa chitinase in dry mature kernels. These two proteins, which have demonstrated antifungal activities both in vitro and in vivo (42, 53, 76), were undetectable or present at low levels in dry mature kernels of susceptible genotypes (42, 71). The presence of high levels of these antifungal proteins in dry kernels of resistant genotypes may significantly delay fungal colonization of corn kernels so that resistant kernels can have enough time to induce an active defense system before it is too late (71).

In addition to antifungal proteins, some storage proteins were also found to be associated with resistance in corn, for example, the 56 kDa GLB1, 46 kDa GLB2, and 24 kDa LEA3 (62, 71). No antifungal activity or clear function has been described for these proteins. However, evidence suggests that globulins may be related to kernel resistance (71, 77-79).

It is worth noting that the synthesis of both GLB1, and LEA3 are modulated by ABA (80). GLB1 accumulation begins early in the maturation phase and specifically requires high levels of ABA. LEA3 accumulation also is dependent upon ABA, but it accumulates much later in embryogenesis, coincident with the onset of dehydration (80), when significant aflatoxin production is usually observed in infected kernels (1). Another unique feature about GLB1 and LEA3 is that they are highly hydrophilic and contain high levels of glycine (>6%) (81). Recently, an aldose reductase, which may be involved in sorbitol biosynthesis and osmoregulation in kernel embryos (72), was also found upregulated in MP420 comparing to susceptible genotypes (71). These discoveries suggest that some of the proteins associated with host resistance in corn against *A. flavus* may also be related to maintenance of proper water potential inside kernels and drought tolerance.

Subpericarp Resistance---Inducible Proteins

Guo *et al.* (24) found that kernels of susceptible genotype increased their resistance significantly when they were imbibed in water for 3 d under germination conditions (31 °C and 100% humidity) to facilitate protein induction prior to fungal inoculation with *A. flavus*. Examination of kernel protein changes during a 7-d incubation under germination conditions found that the 28 kDa chitinase protein, which was undetectable or present at low levels in dry mature susceptible kernels, increased during incubation, becoming detectable after 3 days in susceptible genotypes (71). Other proteins that were found to increase upon fungal infection include beta-1,3-glucanases (30, 35), and some PR proteins (82). These data suggested that induced resistance during germination may be related to increased levels of antifungal proteins in kernels.

Chen *et al.* (71) suggested that water potential may be the primary determinant as to whether a kernel can immediately respond to fungal infection; similar reductions of aflatoxin levels were obtained in kernels of susceptible genotypes imbibed at 0 °C in their study, as were obtained in the previous study (24). However, when subjected to wounding the embryo (83) or freeze-thaw treatment (71) before inoculation, resistant kernels lost their resistance. These studies suggest that de novo synthesis of new proteins by the embryo plays an important role in conferring kernel resistance to *A. flavus*. In reality, however, without high levels of constitutive antifungal proteins as a passive, pre-existing defense mechanism, kernels of susceptible genotypes are likely colonized by *A. flavus* and contaminated with aflatoxins before inducible mechanisms can be established. The real function, therefore, of the high levels of constitutive antifungal proteins may be to delay fungal invasion, and subsequent aflatoxin formation, until new antifungal proteins can be synthesized to form an active defense system. Here, a model of multilayer kernel defense mechanisms is proposed (**Figure 3**).

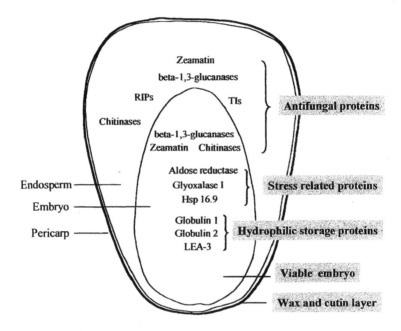

Figure 3. A model of current understanding of host resistance mechanisms in corn against *Aspergillus flavus* infection and/ or aflatoxin production. The level of constitutive and inducible antifungal proteins, some hydrophilic storage proteins and stress related proteins, living embryos, and physical barriers are believe important for host resistance. LEA-3, group 3 late embryogenesis abundant protein; RIP, ribosomal-inactivating protein; TI, trypsin inhibitor.

Genetic engineering of crops for enhanced resistance to aflatoxin contamination using antifungal genes

Currently, the most widely explored strategy to eliminate aflatoxin in corn is to develop preharvest host resistance, through either breeding or genetic engineering since *A. flavus* infects many crops such as corn, cotton, peanut and treenuts, prior to harvest (84). Presently, no practical level of resistance exists in commercial lines of corn and cotton for prevention of attack by *A. flavus* and subsequent aflatoxin contamination. Also, a host resistance strategy may be the easiest strategy for the grower to integrate into the various farming systems. In corn, the host resistance strategy has also gained prominence because of advances in the identification of natural resistance traits (15, 16). It appears that resistance to *A. flavus* infection consists of an interaction of multiple components and biochemical changes that are either preformed or induced upon invasion (15, 62). Genetic studies also have identified multiple chromosome regions associated with resistance to *A. flavus* infection or inhibition of aflatoxin production by restriction fragment length polymorphism (RFLP) analysis (85, 86), suggesting that the resistance trait is quantitatively inherited. Due to the nature of multigene controlled resistance, it is difficulty to move resistance from resistant inbred lines into commercial varieties with desirable agronomic characteristics.

In preliminary studies, several types of genes have demonstrated potential to enhance disease resistance in tobacco or other easily transformed plants (87-89).

Candidate Genes from Corn

Corn has provided several genes that could be significant to resistance against *A. flavus* infection/ aflatoxin contamination. These include genes for antifungal proteins, such as chitinases, beta-1,3-glucanase, 22 kDa TI/zeamatin/thaumatin-like protein, 14 kDa TI, RIP, etc. The 14 kDa TI (42) shown to be correlated with kernel resistance to *A. flavus* infection of corn, when expressed in transgenic tobacco, greatly enhances resistance to the tobacco pathogen, *Colletotrichum destructivum* (87, 88). In addition, leaf extracts from transgenic tobacco inhibited the growth of other phytopathogens such as *Verticillium dahliae*. Cotton is presently being transformed to express the 14 kDa trypsin inhibitor (87). This protein has also been successfully transformed into corn embryonic cultures of A188×B73 and the regenerated transgenic plants are currently being evaluated for resistance against *A. flavus* as well as *F. graminearum* (90). The maize RIP was recently expressed in transgenic rice (89). Whether or not it confers increased resistance in transgenic rice is yet to be determined. Currently, very little literature is available about enhanced host resistance of transgenic plants expressing other maize antifungal genes. However, many studies of transgenic plants expressing other cereal chitinase, beta-1,3-glucanase, or RIP genes have reported increased disease resistance to pathogenic fungi, such as rice blast pathogen *Magnaporthe grisea*, powdery mildew-causing fungus *Erysiphe graminis*, and soilborne fungal pathogen *Rhizoctonia solani* (36, 91, 92).

Candidate Genes from Other Sources

Several recent studies have suggested the potential of manipulating/inducing the lipoxygenase (LOX) pathway in plants to ward off fungal attack, although the potential benefit of enhanced resistance must be balanced against possible adverse effects of LOX products on grain quality (93). The LOX products such as 13-hydroperoxylinoleic acid and its breakdown products/volatiles such as hexenal and hexanal are antifungal and interfere with the aflatoxin pathway. Recently, jasmonic acid, a LOX decay product, was shown to inhibit aflatoxin production and delay spore germination of *A. flavus* (94). The anti-*A. flavus* properties of small chain alkanals and alkenals (derived from the LOX pathway) produced by cotton leaves have also been demonstrated (95, 96). Genes encoding LOX have been identified and cloned from several plant sources (97, 98).

Genes coding for haloperoxidases are also available for possible genetic engineering of plants for antifungal resistance (99, 100), and their utility for plant disease resistance has been documented in transgenic tobacco plants (101, 102). A bacterial chloroperoxidase also greatly reduced the viability of *A. flavus* conidia (100, 102). Leaf extracts from the transgenic tobacco plants were shown to be lethal to germinated spores of *A. flavus* (102).

Certain small lytic peptides have demonstrated convincing inhibitory activity against *A. flavus* and showed promise for transformation of plants to reduce infection of seed. D4E1, a synthetic 17 AA lytic peptide, was shown to interact with sterols present in the conidial cell walls and resist degradation by fungal and host proteases in vitro studies (103). Recently, it was reported that D4E1 gene when transformed into tobacco greatly enhances resistance to *C. destructivum* (88, 104). Treatment of germinating *A. flavus* spores with tobacco leaf extracts from plants transformed with the D4E1 gene significantly reduced spore viability comparing to using extracts from control plants. Preliminary tests of cottonseed transformed with the D4E1 gene also demonstrated enhanced resistance to *A. flavus* penetration in cotton seeds (76).

Conclusion

The control of aflatoxin contamination, especially of corn, will likely be the result of resistant germplasm, developed either by marker-assisted breeding strategies or by genetic engineering of plants with genes expressing resistance against the fungus and/or inhibition of aflatoxin biosynthesis. Naturally resistant corn germplasm not only provides us with a source of resistance, but also nature's lesson concerning specific requirements for the expression of resistance (e.g. antifungal protein compounds, regulation of these factors and physiological conditions for bioactivity). Recent progress in identifying proteins associated with host resistance in corn using a proteomics approach have significantly advanced our knowledge of how corn kernels respond and defend themselves against fungal infection. This may facilitate our future efforts to combine or pyramid multi-resistance genes into susceptible crops for efficient elimination of aflatoxin contamination. Recent progress towards developing preharvest host resistance to aflatoxin contamination of corn may set the stage for the control of other mycotoxin contamination problems as well.

References

1. Payne, G. A. 1998. *In Mycotoxins in Agriculture and Food Safety;* Sinha, K.K.; Bhatnagar, D., Eds.; Marcel Dekker, Inc.: New York, NY, 1998; pp 279-306.
2. Diener, U. L.; Cole, R. J.; Sanders, T. H.; Payne, G. A.; Lee, L. S.; Klich, M. A. *Annu. Rev. Phytopathol.* **1987,** 25, 249-270.
3. Squire, R. A. *Science* **1981,** 214, 877-880.
4. Smith, J. E.; Moss, M. O. *Mycotoxins: Formation Analyses and Significance*; John Wiley and Sons, Chichester, NY, 1985.
5. Hsieh, D. P. H. *In Mycotoxins and Phycotoxins;* Natori, S.; Hashimoto, K.; Ueno, Y., Eds.; Elsevier: Amsterdam, 1989; pp 69-80.
6. Nichols, T. E., Jr. *South. Coop. Ser. Bull.* **1983,** 279, 67-71.
7. Cleveland, T. E.; Bhatnagar, D. *In Molecular Approaches to Improving Food Quality and Safety;* Bhatnagar, D.; Cleveland, T. E., Eds; Van Nostrand Reinhold: New York, NY, 1992; pp 205-228.
8. Klich, M. A.; Tiffany, L. H.; Knaphus, G. *In Aspergillus Biology and Industrial Applications;* Bennett, J. W.; Klich, M. A., Eds.; Butterworth-Heineman: Boston, MA, 1994; pp 329-353.
9. Wicklow, D. T. *Iowa State Univ. Res. Bull.* **1991,** 599, 315-328.
10. Windham, G. L.; William, W. P. *Plant Dis.* **1998,** 82, 281-284.
11. Haumann, F. *Inform* **1995,** 6, 248-256.
12. Brown, R. L.; Cleveland, T. E.; Payne, G. A.; Woloshuk, C. P.; Campbell, K. W.; White, D. G. *Phytopathology* **1995,** 85, 983-989.
13. Lisker, N.; Lillehoj, E. B. In *Mycotoxins and Animal Foods*; Smith, J. E.; Henderson, R. S., Eds.; CRC Press, Inc.: Boca Raton, FL, 1991; pp 689-719.
14. Payne, G. A.; Cassel, D. K.; Adkins, C. R. *Phytopathology* **1986,** 76, 679-684.

15. Brown, R. L.; Cleveland, T. E.; Bhatnagar, D.; Cary, J. E. *In Mycotoxins in Agriculture and Food Safety;* Sinha, K.K.; Bhatnagar, D., Eds.; Marcel Dekker, Inc.: New York, NY, 1998; pp 351-379.

16. Brown, R. L.; Chen, Z. -Y.; Cleveland, T. E.; Russin, J. S. *Phytopathology* **1999,** 89, 113-117.

17. Gardner, C. A. C.; Darrah, L. L.; Zuber, M. S.; Wallin, J. R. *Plant Dis.* **1987,** 71, 426-429.

18. King, S. B.; Scott, G. E. *Phytopathology* **1982,** 72, 942.

19. Widstrom, N. W. In *Aflatoxin in maize: A proceedings of the workshop;* Zuber, M. S.; Lillehoj, E. B.; Renfro, B. L.; Eds.; CIMMYT, El Batan, Mexico, D. F. 1987; pp 212-220.

20. Zuber, M. S.; Darrah, L. L.; Lillehoj, E. B.; Josephson, L. M.; Manwiller, A.; Scott, G. E.; Gudauskas, R. T.; Horner, E. S.; Widstrom, N. W.; Thompson, D. L.; Bockholt, A. J.; Brewbaker, J. L. *Plant Dis.* **1983,** 67, 185-187.

21. Scott, G. E.; Zummo, N. *Crop Sci.* **1988,** 28, 505-507.

22. Campbell, K. W.; White, D. G. *Plant Dis.* **1995,** 79, 1039-1045.

23. Guo, B. Z.; Russin, J. S.; Cleveland, T. E.; Brown, R. L.; Widstrom, N. W. *J. Food Prot.* **1995,** 58, 296-300.

24. Guo, B. Z.; Russin, J. S.; Cleveland, T. E.; Brown, R. L.; Widstrom, N. W. *J. Food Prot.* **1996,** 59, 276-281.

25. Boller, T.; Gehri, A.; Mauch, F.; Vogeli, U. *Planta* **1983,** 157, 22-31.

26. Molano, J.; Polacheck, I.; Duran, A.; Cabib, E. *J. Biol. Chem.* **1979,** 254: 4901-4907.

27. Roberts, W. K.; Selitrennikoff, C. P. *Biochim. Biophys. Acta* **1986,** 880, 161-170.

28. Roberts, W. K.; Selitrennikoff, C. P. *J. Gen. Microbiol.* **1988,** 134, 169-176.

29. Huynh, Q. K.; Hironaka, C. M.; Levine, E. B.; Smith, C. E.; Borgmeyer, J. R.; Shah, D. M. *J. Biol. Chem.* **1992,** 267, 6635-6640.

30. Wu, S.; Kriz, A. L.; Widholm, J. M. *Plant Physiol.* **1994,** 105, 1097-1105.

31. Moore, K. G.; White, D. G.; Payne, G. A. Proceedings of the USDA-ARS Aflatoxin Elimination Workshop (St Louis, MO), 1998, pp 50.

32. Darnetty; Leslie, J. F.; Muthukrishnan, S.; Swegle, M.; Vigers, A. J.; Selitrennikoff, C. P. *Physiol. Plant.* **1993,** 88, 339-349.

33. Mauch, F.; Mauch-Mani, B.; Boller, T. *Plant Physiol.* **1988,** 88, 936-942.

34. Cordero, M. J.; Raventos, D.; San Segundo, B. *Mol. Plant-Microb. Interact.* **1994,** 7, 23-31.

35. Lozovaya, V. V.; Waranyuwat, A.; Widholm, J. M. *Crop Sci.* **1998,** 38, 1255-1260.

36. Jach, G.; Görnhardt, B.; Mundy, J.; Logemann, J.; Pinsdorf, E.; Leah, R.; Schell, J.; Mass, C. *Plant J.* **1995,** 8, 97-109.

37. Richardson, M. *Methods Plant Biochem.* **1991,** 5, 259-305.

38. Chen, M.-S.; Feng, G.; Zen, K. C.; Richardson, M.; Valdes-Rodriguez, S.; Reeck, G. R.; Kramer, K. J. *Insect Biochem. Mol. Biol.* **1992**, 22, 261-268.
39. Richardson, M.; Valdes-Rodriguez, S.; Blanco-Labra, A. *Nature* **1987**, 327, 432-434.
40. Terras, F. R. G.; Schoofs, H. M. E.; Thevissen, K.; Osborn, R. W.; Vanderleyden, J.; Cammue, B. P. A.; Broekaert, W. F. *Plant Physiol.* **1993**, 103, 1311-1319.
41. Halim, A. H.; Wassom, C. E.; Mitchell, H. L.; Edmunds, L. K. *J. Agr. Food Chem.* **1973**, 21, 1118-1119.
42. Chen, Z. -Y.; Brown, R. L.; Lax, A. R.; Guo, B. Z.; Cleveland, T. E.; Russin, J. S. *Phytopathology* **1998**, 88, 276-281.
43. Lorito, M.; Broadway, R. M.; Hayes, C. K.; Woo, S. L.; Noviello, C.; Williams, D. L.; Harman. G. E. *Mol. Plant-Microbe Interact.* **1994**, 7, 525-527.
44. Joshi, B. N.; Sainani,M. N.; Bastawade, K. B.; Gupta, V. S.; Ranjekar, P. K. *Biochem. Biophys. Res. Commun.* **1998**, 246, 382-387.
45. Halim, A. H.; Wassom, C. E.; Mitchell, H. L. *Crop Sci.* **1973**, 13, 405-407.
46. Hochstrasser, K.; Illchmann, K.; Werle E. *Hoppe Seyler's Z. Physiol. Chem.* **1970**, 351, 721-728.
47. Edens, L.; Heslinga, L.; Klok, R.; Ledeboer, A. M.; Maat, J.; Toonen, M. Y.; Visser, C.; Verrips, C. T. Gene **1982**, 18: 1-12.
48. Huynh, Q. K.; Borgmeyer, J. R.; Zobel, J. F. *Biochem. Biophys. Res. Commun.* **1992**, 182, 1-5.
49. Roberts, W. K.; Selitrennikoff, C. P. *J. Gen. Microbiol.* **1990**, 136, 1771-1778.
50. Guo, B. Z.; Chen, Z.-Y.; Brown, R. L.; Lax, A. R.; Cleveland, T. E.; Russin, J. S.; Mehta, A. D.; Selitrennikoff, C. P.; Widstrom, N. W. *Phytopathology* **1997**, 87, 1174-1178.
51. Mahoney, W. C.; Hermodson, M. A.; Jones, B.; Powers, D. D.; Corfman, R. S.; Reeck, G. R. *J. Biol. Chem.* **1984**, 259, 8412-8416.
52. Blanco-Labra, A.; Chagolla-Lopez, A.; Martinez-Gallardo, N.; Valdes-Rodriguez, S. *J. Food Biochem.* **1995**, 19, 27-41.
53. Chen, Z.-Y.; Brown, R. L.; Lax, A. R.; Cleveland, T. E; Russin, J. S. *Appl. Environ. Microbiol.* **1999**, 65, 1320-1324.
54. Hojima, Y.; Pierce, J. V.; Pisano, J. J. *Thromb. Res.* **1980**, 20, 149-162.
55. Walsh, T. A.; Morgan, A. E.; Hey, T. D. *J. Biol. Chem.* **1991**, 266, 23422-23427.
56. Hey, T. D.; Hartley, M.; Walsh, T. A. *Plant Physiol.* **1995**, 107, 1323-1332.
57. Bass, H. W.; Webster, C.; O'Brian, G. R.; Roberts, J. K. M.; Boston, R. S. *Plant Cell* **1992**, 4, 225-234.

58. Linthorst, H. J. M. *Crit. Rev. in Plant Sci.* **1991**, 10, 123-150.
59. Morris, S. W.; Vernooij, B.; Titatarn, S.; Starrett, M.; Thomas, S.; Wiltse, C. C.; Frederiksen, R. A.; Bhandhufalck, A.; Hulbert, S. *Mol. Plant-Microb. Interact.* **1998**, 11, 643-658.
60. Raventós, D.; Cordero, M. J.; San Segundo, B. *Physiol. Mol. Plant Pathol.* **1994**, 45, 349-358.
61. Chen, Z. -Y.; Brown, R. L.; Russin, J. S.; Lax, A. R.; Cleveland, T. E. *Phytopathology* **1999**, 89:902-907.
62. Chen, Z. -Y.; Brown, R. L.; Damann, K. E.; Cleveland, T. E. *Phytopathology* **2000**, 90: S14.
63. Huang, Z.; White, D. G.; Payne, G. A. *Phytopathology* **1997**, 87, 622-627.
64. Wilkins, M. R.; Pasquali, C.; Appel, R. D.; Ou, K.; Golaz, O.; Sanchez, J.-C.; Yan, J. X.; Gooley, A. A.; Hughes, G.; Humphery-Smith, I.; Williams, K. L.; Hochstrasser, D. F. *Biotechnology* **1996**, 14, 61-65.
65. O'Farrell, P. H. *J. Bio. Chem.* **1975**, 250, 4007-4021.
66. Damerval, C.; de Vienne, D.; Zivy, M.; Thiellement, H. *Electrophoresis* **1986**, 7, 52-54.
67. Görg, A.; Boguth, G.; Obermaier, C.; Weiss, W. *Electrophoresis* **1998**, 19, 1516-1519.
68. Appel, R. D.; Palagi, P. M.; Walther, D.; Vargas, J. D.; Sanchez, J. C.; Ravier, F.; Pasquali, C; Hochstrasser, D. F. *Electrophoresis* **1997**,18: 2724-2734.
69. Riccardi, F.; Gazeau, P.; de Vienne, D.; Zivy, M. *Plant Physiol.* **1998**, 117, 1253-1263.
70. Santoni, V.; Bellini, C.; Caboche, M. *Planta* **1994**, 192, 557-566
71. Chen, Z. -Y.; Brown, R. L.; Damann, K. E.; Cleveland, T. E. *Phytopathology* **2001**, unpublished results.
72. Bartels, D.; Engelhardt, K.; Roncarati, R.; Schneider, K.; Rotter, M.; Salamini, F. *EMBO J.* **1991**, 10, 1037-1043.
73. Russin J. S.; Guo, B. Z.; Tubajika, K. M.; Brown, R. L.; Cleveland, T. E.; Widstrom, N. W. *Phytopathology* **1997**, 87, 529-533.
74. Gembeh, S. V.; Brown, R. L.; Grimm, C.; Cleveland, T. E. *Phytopathology* **2000**, 90, S27.
75. Delected
76. Rajasekaran, K.; Cary, J. W.; Jacks, T. J.; Stromberg, K.; Cleveland, T. E. Proceedings of the USDA-ARS Aflatoxin elimination Workshop (Atlanta, GA), 1999, pp. 64.
77. Kriz, A. L. 1989. Biochem. Genet. 27, 239-251.
78. Kriz, A. L.; Wallace, N. H. *Biochem. Genet.* **1991**, 29, 241-254.
79. O'Donoughue, L. S.; Chong, J.; Wight, C. P.; Fedak, G.; Molnar, S. J. *Phytopathology* **1996**, 86, 719-727.

80. Thomann, E. B.; Sollinger, J.; White, C.; Rivin, C. J. *Plant Physiol.* **1992,** 99, 607-614.

81. Garay-Arroyo, A.; Colmenero-Flores, J. M.; Garciarrubio, A.; Covarrubias, A. A. *J. Biol. Chem.* **2000,** 275, 5668-5674.

82. Cordero, M. J.; Raventos, D.; San Segundo, B. *Physiol. Mol. Plant Path.* **1992,** 41, 189-200.

83. Brown, R. L.; Cotty, P. J.; Cleveland, T. E.; Widstrom, N. W. *J. Food Prot.* **1993,** 56, 967-971.

84. Lillehoj, E. B., Proceedings of the CIMMYT aflatoxin in maize workshop (El Batan, Mexico, DF), 1987, pp. 13-32.

85. Davis, G. L.; Williams, W. P. *Maize Genetics Conference* **1999,** 41, 22.

86. Campbell, K. W.; White, D. G. *Phytopathology* **1995,** 85, 886-896.

87. Cary, J. W.; Rajasekaran, K.; Delucca, A. J.; Jacks, T. J.; Lax, A. R.; Cleveland, T. E.; Chlan, C.; Jaynes, J. Proceedings of the USDA-ARS Aflatoxin Elimination Workshop (Memphis, TN), 1997, pp. 55.

88. Rajasekaran, K.; Cary, J. W.; Delucca, A. J.; Jacks, T. J.; Lax, A. R.; Cleveland, T. E.; Chen, Z.; Chlan, C.; Jaynes, J. Proceedings of the USDA-ARS Aflatoxin Elimination Workshop (Memphis, TN), 1997, pp. 66.

89. Kim, J. K.; Duan, X.; Wu, R.; Seok, S. J.; Boston, R. S.; Jang, I. C.; Eun, M. Y.; Nahm, B. H. *Mol. Breed.* **1999,** 5, 85-94.

90. Simmonds, J.; Cass, L.; Lachance, J. Proceedings of the joint meeting of the 50th annual Southern Corn Improvement Conference and the fifty-third annual Northeastern Corn Improvement Conference (Blacksburg, VA), 1998, p. 10.

91. Nishizawa, Y.; Nishio, Z.; Nakazono, K.; Soma, M.; Nakajima, E.; Ugaki, M.; Hibi, T. *Theor. Appl. Genet.* **1999,** 99, 383-390.

92. Bliffeld, M.; Mundy, J.; Potrykus, I.; Futterer, J. *Theor. Appl. Genet.* **1999,** 98, 1079-1086.

93. Doehlert, D. C.; Wicklow, D. T.; Gardner, H. W. *Phytopathology* **1993,** 83, 1473-1477.

94. Goodrich, T. M.; Mahoney, N. E.; Rodriguez S. B. *Microbiol.* **1995,** 141, 2831-2837.

95. Zeringue, H. J., Jr.; McCormick, S. R. *Toxicon.* **1990,** 28, 445-448.

96. Payne, G. A. Proceedings of the USDA-ARS Aflatoxin Elimination Workshop (Memphis, TN), 1997, pp. 66-67.

97. Bell, E.; Mullet, J. E. *Mol. Gen. Genet.* **1991,** 230, 456-462.

98. Melan, M.; Dong, X.; Endara, M. E.; Davis, K. R.; Ausubel, F. M.; Peterman, T. K. *Plant Physiol.* **1993,** 101, 441-450.

99. Wolffram, C.; van Pee, K.-H.; Lingens, F. *FEBS Lett.* **1988,** 238, 325-328.

100. Jacks, T. J.; Delucca, A. J.; Morris, N. M. *Mol. Cell Biochem.* **1999,** 195, 169-172.

101. Jacks, T. J.; Cotty, P. J.; Hinojosa, O. *Biochem. Biophys. Res. Commun.* **1991,** 178, 1202-1204.
102. Rajasekaran, K.; Cary, J. W.; Jacks, T. J.; Stromberg, K.; Cleveland, T. E. *Plant Cell Rep.* **2000,** 19, 333-338.
103. DeLucca, A. J.; Bland, J. M.; Grimm, C.; Jacks, T. J.; Cary, J. W.; Jaynes, J. M.; Cleveland, T. E.; Walsh, T. J. *Can. J. Microbiol.* **1998,** 44, 514-520.
104. Cary, J. W.; Rajasekaran, K.; Jaynes, J. M.; Cleveland, T. E. *Plant Sci.* **2000,** 153, 171-180.

Chapter 12

Reduction of Aflatoxin Contamination in Peanut: A Genetic Engineering Approach

P. Ozias-Akins[1], H. Yang[1], R. Gill[1], H. Fan[1], and R. E. Lynch[2]

[1]Department of Horticulture, The University of Georgia Tifton Campus, Tifton, GA 31793–0748
[2]Crop Protection and Management Research Unit, Agricultural Research Service, U.S. Department of Agriculture, Coastal Plain Experiment Station, Tifton, GA 31793

Development of methods for the introduction of foreign genes into peanut provides an adjunct means to conventional breeding for genetic improvement of the crop for disease resistance. Transformation of peanut is based on microprojectile bombardment of repetitive embryogenic tissue cultures. These cultures can be initiated most efficiently from immature cotyledons or mature embryo axes by culture of the explants on auxin (picloram)-supplemented media. Somatic embryos developing from the primary cultures will undergo repetitive growth when maintained on picloram. Removing auxin from the medium and adding a cytokinin will promote the development of shoots from the somatic embryos. Transformation is accomplished by bombardment of embryogenic cultures with DNA-coated gold particles and selection of transgenic lines on the antibiotic hygromycin. Fewer than 5% of the plants recovered from hygromycin-resistant lines are escapes from selection. Although transformation of peanut by this method is slow, taking approximately 12-14 months, it is highly reproducible and genotype-independent. Aflatoxin contamination of peanut seeds originates with the contaminating fungus, *Aspergillus flavus*, which is an opportunistic saprophyte. Using genetic engineering, we have initiated a three-tiered approach to reduce 1) access of the fungus to the peanut pod, 2) fungal growth, and 3) aflatoxin biosynthesis. This approach encompasses the introduction of insect and fungal resistance genes, as well as genes whose products may interfere with aflatoxin production.

Peanut (*Arachis hypogaea* L.) is an important source of oil and is widely used in the confectionary industry, in candies and in peanut butter. However, unlike soybean (*Glycine max*) which in the US is grown on over 70 million acres, peanut is cultivated on only 1.5 million acres (1). Due to its status as a minor crop in the US, peanut has not received input comparable to soybean for the development of improved cultivars through the application of biotechnology.

Although there are many potential targets for the improvement of peanut by genetic engineering, most notably disease resistance, pest control, and oil composition, one of the most serious industry-wide problems is aflatoxin contamination of peanut seeds which are used in food and feed products. Aflatoxin is a mycotoxin that is produced by *Aspergillus flavus* and *A. parasiticus*, two fungal species that are prevalent in soils. Because peanut pods develop underground, *Aspergillus* can easily invade visibly damaged pods (damaged by insects or mechanical means), but it also may be found infrequently in apparently undamaged pods. Production of aflatoxin, a secondary metabolite in the saprophytic fungus, is enhanced during plant stress induced by drought and high soil temperatures. Aflatoxin has been identified as a carcinogen; therefore, levels in food products are restricted by the Food and Drug Administration to 20 ppb except for milk which has an action level of 0.5 ppb. These regulations require that aflatoxigenic fungi and aflatoxin levels be monitored from the buying point through peanut processing.

Approaches to control preharvest aflatoxin contamination range from biocontrol with atoxigenic strains, to modified cultural practices and the introduction of genetic resistance (2). Identification of quantitative genetic resistance within the gene pool holds promise, although screening methods require extensive replication to produce meaningful results (3). In addition to accessing potential resistance from the *A. hypogaea* gene pool, genetic engineering to introduce genes for fungal resistance, insect resistance (since aflatoxin levels are correlated with insect damage), or reduction of aflatoxin biosynthesis may be a realistic adjunct to traditional breeding. Our goal has been to develop a reproducible, genotype-independent transformation system for peanut that can be used to test the efficacy of foreign genes for reducing aflatoxin contamination.

A Reproducible, Genotype-Independent Transformation System for Peanut

Most transformation systems, whether mediated by biological means (*Agrobacterium tumefaciens*) or by physical methods of free DNA uptake (microprojectile bombardment, electroporation, silica carbide whiskers, etc.), rely on efficient plant regeneration from tissue cultures of the species of interest (*4*). In peanut, dedifferentiation of tissues into a true callus phase (unorganized growth and cell division) eliminates regeneration ability. A variety of immature tissues can, however, be induced to directly form shoot primordia or somatic embryos when plated on the appropriate growth regulators. In our hands, induction of somatic embryos from immature cotyledons or embryo axes of peanut seeds can be accomplished with all genotypes tested when the explant tissues are plated on a standard tissue culture medium (*5*) that has been supplemented with the growth regulator, picloram (*6-8*; Fig. 1). Somatic embryos form directly from the explanted tissue without an intervening callus phase. The embryogenic growth typically originates from the portion of the cotyledon proximal to the cotyledonary node or from the epicotyledonary portion of the embryo axis either between the first true leaves and the axillary buds at the cotyledonary node or from the young leaf primordia. The embryogenic tissues are capable of repetitive embryogenesis when subcultured at regular intervals onto medium of the same composition.

Whole peanut plants can be regenerated from mature somatic embryos using a 2- to 3-step protocol (*9*). When root and shoot poles of a somatic embryo do not develop and elongate simultaneously, the shoot can be excised and easily rooted on medium containing 0.2 mg/l naphthaleneacetic acid. Although no tap root is subsequently present to support growth of the plant, numerous adventitious roots allow a rapid transition from agar-based culture medium to potting mix.

A uniform, readily regenerable tissue culture facilitates plant transformation. The embryogenic tissue culture should consist primarily of translucent, smooth-surfaced somatic embryos between globular and early cotyledonary stages of development in order to increase the probability of recovering transgenic cell lines. Foreign genes can be introduced into uniform, embryogenic cultures of peanut by microprojectile bombardment (*9*). We typically bombard cultures (10-14 days after subculture) with DNA-coated gold particles accelerated by bursting of an 1800 psi rupture disc with helium pressure. Just prior to bombardment, tissue pieces are arranged in a 2 cm-diameter circle in the center of the culture dish where they

154

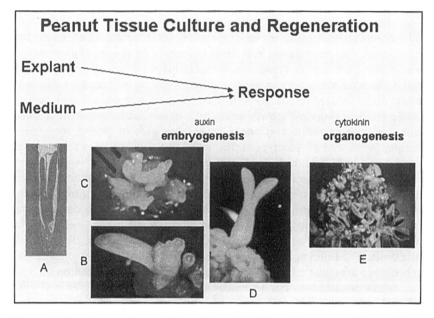

Figure 1. Immature embryos (A) can be excised from the surrounding seedcoat tissue and separated into cotyledons and embryo axis. Cultured cotyledons will form somatic embryos at their base (B) which can be subcultured to produce repetitive embryogenic cultures (C). Removing picloram from the culture medium will allow somatic embryos to develop further (D). Organogenic cultures in which shoots, rather than somatic embryos, develop can be induced on media containing various cytokinins (E). (B and D are reproduced with permission from reference 8. Copyright 1992 Urban and Fischer Verlag)

remain until 2-3 days post-bombardment. If the tissues were bombarded with a construct containing the reporter gene, β-glucuronidase (GUS), pieces of tissue can be stained to assess the effectiveness of the bombardment conditions (Fig. 2). Tissues cultured under non-selective conditions and stained for GUS activity after one month may show the development of a transgenic cell lineage (Fig. 2). Since the GUS assay is destructive, such lineages cannot be selectively transferred, although the use of a non-destructive reporter such as the green fluorescent protein might allow small sectors to be manually recovered (*10*). We have opted for antibiotic selection of stably transformed cells which is most effective with hygromycin. The hygromycin phosphotransferase gene (*hph*) is expressed sufficiently in embryogenic tissues when driven by either the CaMV35S promoter or the potato ubiquitin 3 promoter (*11*). Hygromycin resistance is clearly expressed not only during selection for embryogenic tissues *in vitro*, but also during rooting and in young, mature leaves (Fig. 3).

Regeneration of plants from transgenic cell lines is possible if the cell lines have remained clearly embryogenic. Morphologically normal, flowering plants usually are obtained from transgenic somatic embryos. However, peanut appears to be unusually susceptible to the deleterious effects of the tissue culture and transformation process with regard to flowering and fertility. Although plant regeneration occurs readily from embryogenic cultures even after extended culture duration, many of the primary regenerants display delayed flowering and partial or complete sterility. For this reason, it is recommended that tissues be maintained only as long as necessary to establish a sufficient number of uniform cultures to carry out bombardments within 4-9 months of culture initiation. Fourteen months typically are required to complete the transformation cycle from seed to seed including culture initiation, bombardment and selection, regeneration, acclimitization and maturation. Transformation of peanut using the general process outlined above has been successfully repeated in our lab as well as others (*9, 12-16*).

Application of Peanut Genetic Engineering to the Problem of Aflatoxin Contamination

We are taking a three-tiered approach to reduction of pre-harvest aflatoxin contamination, using genetic engineering, that addresses insect damage, fungal growth and aflatoxin biosynthesis. Only our efforts to control insect damage in peanut have progressed to the point of efficacy testing in the field, thus they will be described in the most detail below. It has been clearly documented that aflatoxin contamination is positively correlated with insect damage (*17*). In peanut, the insect pest most commonly associated with aflatoxin contamination is the lesser cornstalk borer (*Elasmopalpus lignosellus*; LCB). The larvae of LCB often tunnel into

156

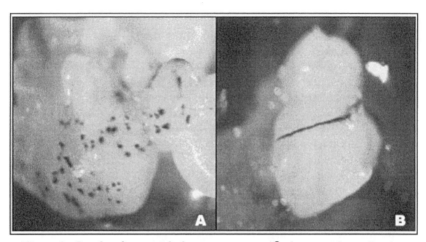

Figure 2. Bombardment with the reporter gene β-glucuronidase allows an assessment of transient expression (A) or formation of stably transformed sectors (B).

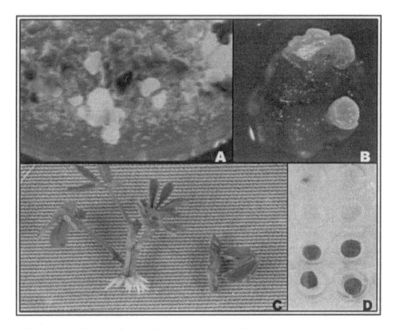

Figure 3. The antibiotic, hygromycin can be used to select transgenic embryogenic tissues in liquid (A) or agar (B) medium, at rooting (C), or during screening of leaf discs from regenerated plants and progeny (D). (B reproduced with permission from reference 9. Copyright 1993 Elsevier Science)

peanut stems and feed on pods developing underground (*18*). Damage to pods can be due to penetration of the pod wall or extensive scarification of the pod surface (Fig. 4). LCB thrives in hot, dry weather when soil temperatures are high, and these also are the most favorable conditions for aflatoxin production. To control the amount of pod damage attributed to LCB, the most effective means may be by host plant resistance since soil insecticides have a variable period of residual activity. LCB is a lepidopteran insect pest and thus susceptible to several of the insecticidal crystalline proteins found in the soil bacterium, *Bacillus thuringiensis* (*19*). Peanut genotypes transformed with Bt *cryIA*(c) are resistant to foliar damage by LCB when the larvae are forced to feed on leaves *in vitro* (*12*; Fig. 4). Field resistance to insect damage to foliage and pods has been observed (Fig. 4). Four transgenic lines recovered by transformation of cultivar MarcI in 1995 have been carried forward during field testing. These lines initially were selected based on *in vitro* assays at the T1 generation (Table 1). One of the lines does not carry an intact Bt *cryIA*(c) gene as shown by PCR analysis of multiple T3 individuals (Fig. 5). *CryIA*(c)-expressing and control lines are being field tested in 2000 for both LCB resistance as well as aflatoxin reduction.

Table I. LCB Bioassay on Transgenic Peanut

Line	Survival (%)	Weight (mg)	Damage (%)
22 (89-5-23)	0	0	<25
24 (89-5-24)	0	0	<25
124 (89-4-2)	100	13	50
137 (89-4-6)	<50	<7	<50

Reducing insect damage is likely to have an impact on pre-harvest aflatoxin contamination of peanut because of the close association between these two pests. However, insect resistance alone probably will not be sufficient for total aflatoxin control. Alternative and supplementary strategies are based on inhibition of fungal growth or mycotoxin production. One putative antifungal gene that we have introduced into peanut is an anionic peroxidase from tomato (*tap1*; *20*). This highly anionic peroxidase localizes at the site of suberization of wound-healing cells. Expression of different peroxidases in transgenic plants has been shown to confer some degree of resistance to selected fungal pathogens and even to enhance insect resistance (*21, 22*). Testing of transgenic peanut lines overexpressing peroxidase for fungal and insect resistance is in progress.

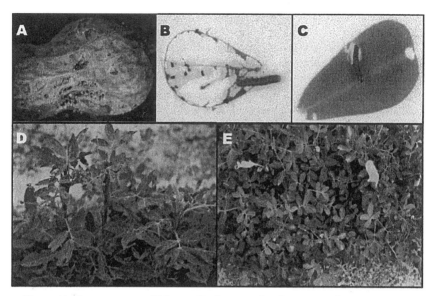

Figure 4. Lesser cornstalk borer feeds on pods of peanut, often extensively scarifying the surface (A). Control leaflets are extensively damaged by lepidopteran insect larvae (B), whereas leaflets from transgenic peanut plants expressing cryIA(c) show little damage (C). Damage by foliar-feeding insects is apparent on the parental cultivar, Marc I (D), but visibly less on transgenic peanut (E).

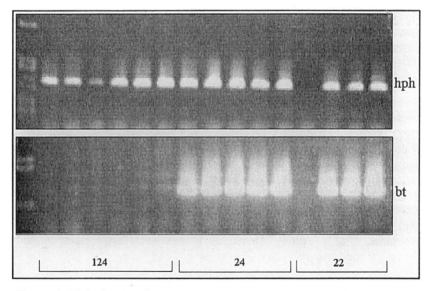

Figure 5. Multiple T3 individuals from three transgenic peanut lines (124, 24, 22) tested by PCR for amplification of a portion of the hygromycin phosphotransferase gene (hph) and the Bt cryIA(c) gene (bt).

In conclusion, preharvest aflatoxin contamination likely cannot be controlled by a single strategy, but will require a combination of genetic resistance and improved cultural practices. Genetic resistance available within the *Arachis* gene pool can be augmented by the introduction of foreign genes that would confer specific traits such as insect resistance, inhibition of fungal growth, or inhibition of aflatoxin biosynthesis. According to our results, expression of CryIA(c) can significantly reduce the damage to peanut pods caused by the lesser cornstalk borer, but the effect on aflatoxin contamination using any of the suggested genetic engineering approaches remains to be tested.

References

1. Http://www.usda.gov/nass/pubs/trackrec
2. Aflatoxin Elimination Workshops, USDA-ARS, National Program in Food Safety
3. Anderson, W. F.; Holbrook, C. C.; Wilson, D. M.; Matheron, M. E. Evaluation of preharvest aflatoxin contamination in several potentially resistant peanut genotypes. *Peanut Sci.* **1995**, *22*, 29-32.
4. Hansen, G.; Wright, M. S. Recent advances in the transformation of plants. *Trends in Plant Science* **1999**, *4*, 226-231.
5. Murashige, T.; Skoog, F. A revised medium for rapid growth and bioassay with tobacco tissue cultures. *Physiol. Plant.* **1962**, *15*, 473-497.
6. Ozias-Akins, P. Plant regeneration from immature embryos of peanut. *Plant Cell Rep.* **1989**, *8*, 217-218.
7. Ozias-Akins, P.; Anderson, W. F.; Holbrook, C. C. Somatic embryogenesis in *Arachis hypogaea* L.: genotype comparison. *Plant Sci.* **1992**, *83*, 103-111.
8. Ozias-Akins, P.; Singsit, C.; Branch, W.D. Interspecific hybrid inviability in crosses of *Arachis hypogaea* x *A. stenosperma* can be overcome by in vitro embryo maturation or somatic embryogenesis. *J. Plant Physiol.* **1992**, *140*, 207-212.
9. Ozias-Akins, P.; Schnall, J. A.; Anderson, W. F.; Singsit, C.; Clemente, T. E.; Adang, M. J.; Weissinger, A. K. Regeneration of transgenic peanut plants from stably transformed embryogenic callus. *Plant Sci.* **1993**, *93*, 185-194.
10. Vain, P.; Worland, B; Kohli, A; Snape, J. W.; Christou, P. The green fluorescent protein (GFP) as a vital screenable marker in rice transformation. *Theor. Appl. Genet.* **1998**, *96*, 164-169.
11. Garbarino, J. E.; Belknap, W. R. Isolation of a ubiquitin-ribosomal protein gene (*ubi3*) from potato and expression of its promoter in transgenic plants. *Plant Molec. Biol.* **1994**, *24*, 119-127.
12. Singsit, C.; Adang, M. J.; Lynch, R. E.; Anderson, W. F.; Wang, A.; Cardineau, G.; Ozias-Akins, P. Expression of a *Bacillus thuringiensis cryIA*(c)

gene in transgenic peanut plants and its efficacy against lesser cornstalk borer. *Transgenic Res.* **1997**, *6*, 169-176.

13. Wang, A.; Fan, H.; Singsit, C.; Ozias-Akins, P. Transformation of peanut with a soybean *vspB* promoter-*uidA* chimeric gene. I. Optimization of a transformation system and analysis of GUS expression in primary transgenic tissues and plants. *Physiol. Plant.* **1998**, *102*, 38-48.

14. Yang, H.; Singsit, C.; Wang, A.; Gonsalves, D.; Ozias-Akins, P. Transgenic peanut plants containing a nucleocapsid protein gene of tomato spotted wilt virus show divergent levels of gene expression. *Plant Cell Rep.* **1998**, *17*, 693-699.

15. Livingstone, D. M.; Birch, R. G. Efficient transformation and regeneration of diverse cultivars of peanut (*Arachis hypogaea* L.) by particle bombardment into embryogenic callus produced from mature seeds. *Molec. Breed.* **1999**, *5*, 43-51.

16. Chenault, K. D.; Burns, J. A.; Melouk, H. A. β-1,3-glucanase activity in transgenic peanut. *Amer. Peanut Res. Educ. Soc. Proc.* **1999**, *31*, 69.

17. Lynch, R. E.; Wilson, D. M. Enhanced infection of peanut, *Arachis hypogaea* L. seeds with *Aspergillus flavus* group fungi due to external scarification of peanut pods by the lesser cornstalk borer, *Elasmopalpus lignosellus* (Zeller). *Peanut Sci.* **1991**, *18*, 110.

18. *Peanut Health Management*; Melouk, H. A.; Shokes, F. M., Eds.;American Phytopathological Society; St. Paul, MN, 1995.

19. Schnepf, E.; Crickmore, N.; Van Rie, J.; Lereclus, D.; Baum, J.; Feitelson, J.; Zeigler, D. R.; Dean, D. H. *Bacillus thuringiensis* and its pesticidal crystal proteins. *Microbiol. Molec. Biol. Rev.* **1998**, *62*, 775.

20. Roberts, E.; Kolattukudy, P. E. Molecular cloning, nucleotide sequence, and abscisic acid induction of a suberization-associated highly anionic peroxidase. *Molec. Gen. Genet.* **1989**, *217*, 223-232.

21. Kazan, K.; Goulter, K. C.; Way, H. M.; Manners, J. M. Expression of a pathogenesis-related peroxidase of *Stylosanthes humilis* in transgenic tobacco and canola and its effect on disease development. *Plant Sci.* **1998**, *136*, 207-217.

22. Dowd, P. F.; Lagrimini, L. M. Examination of different tobacco (*Nicotiana* spp.) types under- and overproducing tobacco anionic peroxidase for their leaf resistance to *Helicoverpa zea. J. Chem. Ecol.* **1997**, *23*, 2357-2370.

Chapter 13

Development of Micropropagation Technologies for St. John's wort (*Hypericum perforaturm L.*): Relevance on Application

S. J. Murch, S. D. S. Chiwocha, and P. K. Saxena

Department of Plant Agriculture, University of Guelph, Guelph, Ontario N1G 2W1, Canada

Introduction

Phytopharmaceuticals are medicinal plant preparations with a long history of anecdotal evidence of efficacy, extensive biochemical characterizations, proven effectiveness in placebo-controlled clinical trials and in some cases, standardization and sale with a Drug Identification Number (DIN). *Hypericum perforatum* (St. John's wort) is a medicinal plant with a long history of use for the treatment of neurological disorders and depression (*1, 2, 3, 4*). In 1998, 7.5 million Americans used St. John's wort for the treatment of neurological disorders and depression (*5*) based on a demonstrated efficacy in numerous clinical trials (*3*). In 1997, the National Institute of Health (NIH), Office of Alternative Medicines, began a 3-year-study costing $4.3 million to compare the effects of *H. perforatum*, a placebo and a standard anti-depressive drug in patients suffering from mild depression (*6*). Regardless of the outcome of the NIH study, it will be necessary to solve several ongoing problems with preparations of St. John's wort before consumers can use this phytopharmaceutical effectively. A recent study found that there was a 17-fold difference in the content of the marker compound hypericin and a 13-fold difference in pseudohypericin content of commercially prepared St. John's wort capsules (*7*).

The popularity of St. John's wort preparations has increased immensely, 2800% in one year (*8*), despite a high degree of variability in product quality and limited available scientific information concerning the unique physiology of medicinal plants. One of the many reasons for this popularity is the public's perception of "natural" as equivalent to "safe". As a result, the fundamental question of why medicinal plants are different from other plant species has been left largely unanswered. The identity of the medicinally active constituents has remained elusive and most preparations are sold on the basis of the concentration of a marker compound. In the case of St. John's wort, this compound is commonly hypericin. However, studies into the unique physiology of St. John's wort have identified more than 25 potentially "active" compounds (*9, 10, 11*) including the recent report of relatively high levels of the mammalian neurohormone melatonin (*12*).

One of the problems associated with medicinal plant preparations is the extreme variability in the content of marker compounds. Variability in the content of endogenous metabolites in several medicinal plant species has been found to occur as a result of the growing conditions and management practices for the crop. The synthesis of the basic skeletons for active secondary metabolites is dependent on the carbon assimilated during photosynthesis. Therefore the photoperiod and the intensity and spectrum of the available light during a cropping season have been shown to influence the medicinal content of plant preparations (*13*). Similarly, warm weather conditions favor the synthesis of secondary metabolites while rainy weather can inhibit alkaloid production in many species (*13*). As well, the availability of nutrients in the soil profoundly affects the chemical composition of the medicinal plant preparations (*13*). While it is difficult to separate the various factors under field conditions, their implications on medicinal plant production can be serious.

A second area of concern has been the reports of adulteration of St. John's wort preparations with misidentified plant species and other *Hypericum* species including *H. maculatum, H. barbatum, H. hirsutum, H. montanum* and *H. tetrapterum* (*14*). As well, St. John's wort has been mistaken for Rose of Sharon (*Hypericum calycinum*) (*15*).

Another lingering problem with the production of St. John's wort has been the lack of chemical methods for analysis and standardization. Recently, researchers at the U.S. Food and Drug Administration's National Center for Toxicological Research have determined the optimal conditions for extracting several active constituents from St. John's wort leaf tissues and for the quantification of the compounds by HPLC (*16*).

In addition, St. John's wort preparations are produced from field-grown crops and therefore are susceptible to infestation by bacteria, fungi, insects and environmental pollutants that can alter the medicinal content of the preparations and compromise the health of consumers (*15*). Therefore, the principal prerequisite for the preparation of high quality phytopharmaceuticals is the identification, quantification and optimization of the conditions for production of the unique metabolites in medicinal plant species.

Antidepressant activity of St. John's wort

As described above, St. John's wort preparations contain a complex balance of more than two dozen bioactive compounds including: naphthodianthrones (hypericin, pseudohypericin, isohypericin, protohypericin), flavonoids (amentoflavone, hyperin, kaempferol, luteolin, myricetin, quercetin), phloroglucinols (hyperforin, adhyperforin, hyperoside, leucocyanidin), antioxidants (proanthocyanidins, procyanidins), tannins, coumarins (umbellifone and scopoletin), xanthones, essential oils, amino acids, organic acids, and carotenoids (11). Thus, the reported beneficial effects of St. John's wort extracts on a wide range of disorders may be due to the interactions of several of these compounds.

Current treatment for depression involves the use of medications that hinder the transport of neurotransmitters, including serotonin and norepinephrine (17). Although extracts of St. John's wort are commonly used for treating mental depression, the precise mechanism governing the antidepressant effect of this phytopharmaceutical is still unknown (17, 18). Reports made recently indicate that *H. perforatum* extracts may act in a manner analogous to synthetic antidepressants by inhibiting the transport of neurotransmitters. The *Hypericum* extract reduced serotonin and norepinephrine re-uptake into astrocytes (17) and decreased the expression of ß-receptors in the frontal cortex in rats (19). Whether the inhibition is due to the activity of one compound or an interaction of several components present in the extract is still under investigation. As was previously pointed out by Cott (4), the standardization of formulations necessitates the identification of the active components required for the curative effect of St. John's wort extracts. Recently, the research efforts of several workers have been focused on addressing this particular problem.

The compound, hyperforin, has been identified to be important for the medicinal effects of St. John's wort as an antidepressant (20, 21). Recently, this compound has also been shown to be the active biomolecule responsible for the antibacterial activity of *Hypericum* extracts (22). A hyperforin-enriched CO_2 extract inhibited the synaptosomal uptake of serotonin, norepinephrine and dopamine, demonstrating that this compound is the major antidepressant constituent of St. John's wort extracts (19). When the compound was administered into rats, hyperforin led to an increase in the extracellular concentration of neurotransmitters in the locus coeruleus (23). Evidence suggesting that the decreased uptake of serotonin is not due to direct binding of

St. John's wort extracts or pure hyperforin with the transporter protein was provided by Gobbi et al. (*18*). As an alternative mechanism, they suggested that interaction of *H. perforatum* extract with storage vesicles for serotonin in synaptosomes leads to an increase in the cytoplasmic concentration of the neurotransmitter, and that it is this increase that causes a reduction in serotonin re-uptake (*18*). Results from another study by Singer et al. (*24*)· suggest that hyperforin is a sodium ionophore that may inhibit serotonin uptake by increasing the intracellular concentration of Na^+. The ability of hyperforin to indiscriminately inhibit the re-uptake of several neurotransmitters (i.e. dopamine, norepinephrine and serotonin) into synaptosomes may be based on this capability to elevate intracellular Na^+ concentrations (*24*). Another compound, furohyperforin, which is a polar analogue of hyperforin, was recently isolated from *Hypericum perforatum* (*25*). However, furohyperforin was not as effective as hyperforin in inhibiting serotonin re-uptake into brain cortical synaptosomes, indicating that this constituent of St. John's wort is not the major antidepressive component (*25*).

It is apparent from the literature that the evidence supporting a role for hyperforin as one of the most important neuroactive compounds in *H. perforatum* extracts responsible for their antidepressant effects is mounting. Chatterjee et al. (*26*) recommended this molecule as a starting point in the development of antidepressant drugs with different mechanisms of action to those of synthetic antidepressants currently available. However, working with this compound presents a few problems. Since pure hyperforin is highly susceptible to oxidation and is also light sensitive, conditions that limit its degradation during long-term storage need to be established (*27*). The degree of degradation of hyperforin in St. John's wort extracts is less than that of the isolated and purified compound, and this is conceivably due to the presence of natural antioxidants in the extracts (*27*). Furthermore, hyperforin was recently identified as the compound that is responsible for the interaction of St. John's wort extracts with other drugs (*28*). The compound binds to the pregnane X receptor (PXR) in human liver cells. This results in the activation of PXR, which in turn promotes the expression of cytochrome P450 3A4, an enzyme that is involved in the metabolism of several drugs in the liver (*28*). Therefore, although hyperforin from St. John's wort is beneficial for the treatment of mental depression, it also has detrimental effects when taken with other drugs that are metabolized by cytochrome P450 3A4 (*28*).

Another interesting finding was the report of mammalian hormones, including serotonin and melatonin in St. John's wort tissues (*12*). Recently, a biosynthetic pathway for serotonin and melatonin that appears to be analogous to the established pathway in mammals, yeast, and bacteria was demonstrated in St. John's wort seedlings (Figure 1). Interestingly, although there is no known role for melatonin in plant morphogenesis or physiology, Balzer and Hardeland (*29*) hypothesized that melatonin in plants may have an analogous

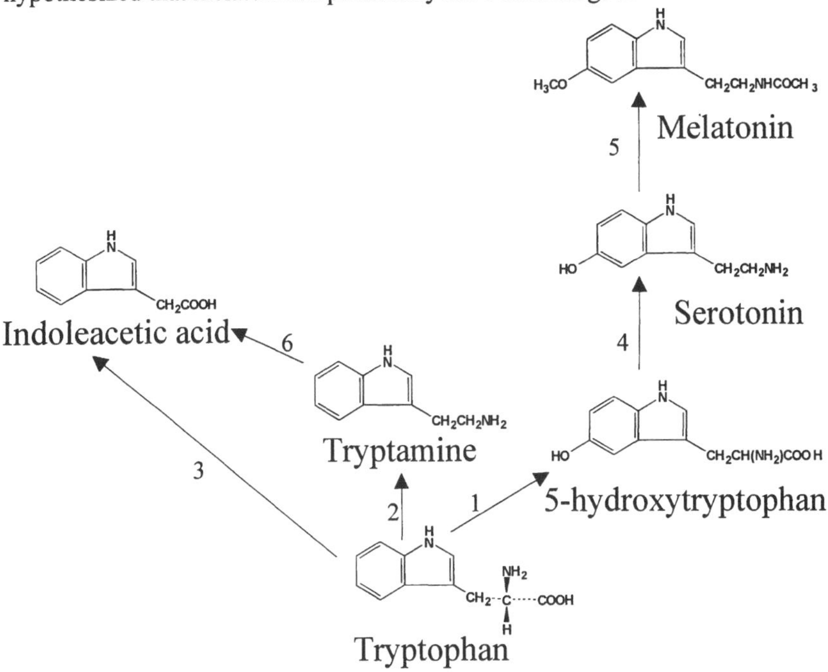

Figure 1. Metabolic pathways of tryptophan in mammals and plants (Reproduced with permission from reference 34. Copyright 2000 Springer). (1) tryptophan 5-hydroxylase EC 1.14.16.4 (2) tryptophan decarboxylase EC 4.1.1.28 (3) L-tryptophan transaminase and decarboxylase EC 1.4.1.19; 4.1.1.43; 1.2.3.7 (4) L-amino acid decarboxylase EC 4.1.1.28 (5) serotonin N-acetyltransferase EC 2.3.1.5 (6) tryptamine deaminase EC 1.13.11.11.

role to that in mammals, acting as a chemical messenger of light and dark, calmodulin binding factor or an antioxidant. In this way, the relative ratios of melatonin and serotonin may be involved in regulation of light:dark responses, seasonality and circadian rhythms in both plants and mammals (*29, 30*).

From all of these studies, it appears that St. John's wort contains a unique and complex biochemical profile and that it is unlikely that a single, medicinally active compound will be identified that can account for the range of effects and responses. On the basis of the presence of this diverse array of potentially bioactive compounds it is logical to assume that the synergistic effects of several compounds within the plant tissues may be responsible for the therapeutic effects.

Development of new technologies

One of the challenges associated with the production of St. John's wort is the optimization of conditions to provide plant materials with an optimal profile of more than 20 biologically active constituents identified to date. Therefore, technologies are required which allow for the large-scale production of optimized, intact plant tissues in a sterile, controlled environment. One solution to the problems faced by the medicinal plants industry is the development of *in vitro* systems for the production of medicinal plants. The production of plants *in vitro* has several advantages: a) plants are grown in sterile, standardized conditions; b) individual superior plants can be identified and clonally propagated; c) plant material is consistent and therefore, precise biochemical characterizations can be achieved; and d) eventually protocols can be developed for the improvement of the crop through genetic manipulation. The implementation of *in vitro* systems for the growth and optimization of medicinal plant species represents the first step in the production of consistent, high-quality phytopharmaceutical preparations.

Micropropagation of St. John's wort

In vitro propagation is the process by which it is possible to generate hundreds of identical plantlets in sterile culture (*31*). The development of an efficient system for the regeneration of St. John's wort plantlets *in vitro* was recently described (*32, 33*). *De novo* shoot regeneration was effectively induced on etiolated hypocotyls and stem segments excised from sterile seedlings by exposure to the plant growth regulator thidiazuron. In these experiments, the optimal concentration and duration of the TDZ exposure were determined to be 5 $\mu mol \bullet L^{-1}$ for 6 or 9 days, respectively (*32, 33*). With this treatment regime, an average of 30-40 new shoots were developed on each 1 cm piece of tissue (Figure 2). Visual observations of the stem segments revealed that the epidermis split and the regenerants originated from the endodermal layers (Figure 2A). Regenerants developed further and red-coloured hypericin glands were visible on the developing shoots (Figure 2B). After an incubation period of 30 days, the entire surface of the tissue was covered with regenerants (Figure 2C). Regenerated shoots were transferred to Magenta boxes where plantlets formed with extensive roots and shoots (Figure 2D). Further development of the

Figure 2. De novo shoot regeneration on stem sections of St. John's wort (Hypericum perforatum L.) (Reproduced with permission from refeence 34. Copyright 2000 Springer). A. De novo shoot organogenesis developing at the split in the epidermis in response to culture on a medium supplemented with thidiazuron for 6 days (Bar 0.03 cm). B. Development of the shoots on the surface of the stem section. Note the appearance of hypericin glands around the margin of the developing leaves (Bar 0.015 cm). C. Clusters of regenerants formed on each stem section by day 30 (Bar 0.6 cm). D. St. John's wort plantlets grown in Magenta boxes for a period of 2 months (Bar 1.2 cm).

plantlets in a bioreactor provided large amounts of sterile consistent plant material for biochemical analysis and commercial production (*33*).

Future Directions

As new technologies are developed for the assessment and manufacture of plant-based medicines, the challenges presented by the plant tissues need to be accommodated. Novel technologies are required that allow for the production of whole plant tissues with complex, optimal biochemical profiles in the absence of abiotic and biotic contamination. The development of technologies for *in vitro* culture of St. John's wort and other medicinal plant species provide the first steps for the eventual production of high-quality plant-based medicines. Initial investigations are required to develop a full profile of the active biomolecules in St. John's wort and other medicinal plant species. This necessitates the development of new protocols for the extraction and quantification of the active compounds from St. John's wort. Additional work will be needed to determine the effects of varying environmental conditions on the production of individual secondary metabolites. The availability of *in vitro* protocols provides the technical basis for these ongoing investigations.

Additionally, medicinal plants have traditionally been wild-harvested with only limited efforts to breed for specific compounds. As a result, harvested St. John's wort from different sources may represent radically different genetic pools, whether superior or inferior. The techniques of plant cell culture and genetic manipulation offer an interesting alternative for broadening the pool of genetic variability in St. John's wort. A number of genes of agricultural significance such as herbicide and pesticide resistance have been transferred to many crops using various gene transfer methods. However, it is expected that because of the complex nature of the biochemical processes involved in the synthesis of the bioactive secondary metabolites, the molecular characterization of genes may take longer than is the case with single gene metabolites. Until the time that all of the individual genes regulating entire, complex pathways are identified, cloned, and packaged into appropriate vectors, transformation technologies cannot easily be applied. Therefore, those approaches that allow for the transfer of bulk DNA, with a known or unknown complement of genes, may not only produce interesting novel germplasm for fundamental studies, but may also lead to the development of plants with synergistic combinations of secondary metabolites for novel treatments. The next phase of this research will be the development of protocols for the application of cell fusion techniques to the genetic improvement of medicinal plants.

References

1. Sommer, H.; Harrer, G. *J. Geriatr. Psychiatry. Neurol.* **1994**, *7*, S9-S11.
2. Hansgen, K.D.; Vesper, J.; Ploch, M. *J. Geriatr. Psychiatry. Neurol.* **1994**, *7*, S15-S18.
3. Linde, K.; Ramirez, G,; Mulrow, C. D.; Pauls, A.; Weidenhammer, W.; Melchart, D. *Br. Med. J.* **1996**, *313*, 253-258.
4. Cott, J. M. *Pharmacopsychiat.* **1997**, *30 (Supp.)*, 108-112.
5. Greenwald, J. *Time.* **1998**, November 23, 1998. 48-58.
6. National Institute of Health. October 1997, *4*, p5.
7. Consumer Safety Symposium on Dietary Supplements and Herbs. Good Housekeeping Institute. New York, NY. March 3, 1998.
8. Ernst, E. *Br. Med. J.* **2000**, *321*, 395-396.
9. Nahrstedt, A.; Butterweck, L. *Pharmacopsychiat.* **1997**, *30*, 129-134.
10. Evans, M. F.; Morgenstern, K. *Can. Fam. Physician.* **1997**, *43*, 1735-1736.
11. Miller, A. L. *Altern. Med. Rev.* **1998**, *3*, 18-26.
12. Murch, S. J.; Simmons, C. B.; Saxena, P. K. *The Lancet.* **1997**, *350*, 1598-1599
13. Bernath, J. *Herbs, Spices and Med. Plants.* **1986**, *1*, 185-234.
14. St. John's Wort Monograph. *HerbalGram.* **1997**, *40*, 37-45.
15. Hobbs, C. *HerbalGram.* **1989**, *18/19*, 24-33.
16. Liu, F. F.; Springer, D. *J. Agric. Food Chem.* **2000**, *48*, 3364.
17. Neary, J. T.; Bu, Y. *Brain Res.* **1999**, *816*, 358-363.
18. Gobbi, M.; Valle, F. D.; Ciapparelli, C.; Diomede, L.; Morazonni, P.; Verotta, L.; Caccia, S.; Cervo, L.; Mennini, T. *Naunyn-Schmiedeberg's Arch. Pharmacol.* **1999**, *360*, 262-269.
19. Müller, W. E.; Singer, A.; Wonnemann, M.; Hafner, U.; Rolli, M.; Schäfer, C. *Pharmacopsychiat.* **1998**, *31(Suppl.)*, 16-21.
20. Laakmann, G.; Schule, C.; Baghai, T.; Kieser, M. *Pharmacopsychiat.* **1998**, *31*, 54-9.
21. Chatterjee, S. S.; Nöldner, M.; Koch, E.; Erdelmeier, C. *Pharmacopsychiat.* **1998**, *31(Suppl.)*, 7-15.
22. Schempp, C. M.; Pelz, K.; Wittmer, A.; Schöpf, E.; Simon, J. C. *The Lancet.* **1999**, *353*, 2129.
23. Kaehler, S. T.; Sinner, C.; Chatterjee, S. S.; Philipu, A. *Neurosci. Letters.* **1999**, *262*, 199-202.
24. Singer, A.; Wonnemann, M.; Müller, W. E. *J. Pharm. Exp. Therapeutics.* **1999**, *290*, 1363-1368.

25. Verotta, L.; Appendino, G.; Belloro, E.; Jakupovic, J.; Bombardelli, E. *J. Nat. Prod.* **1999,** *62,* 770-772.
26. Chatterjee, S. S.; Bhattacharya, S. K.; Wonnemann, M.; Singer, A.; Müller, W. E. **1998,** *Life Sci. 63,* 499-510.
27. Orth, H. C. J.; Rentel, C.; Schmidt, P. C. *J. Pharm. Pharmacol.* **1999,** *51,* 193-200.
28. Moore, L. B.; Goodwin, B.; Jones, S. A.; Wisely, G. B.; Serabjit-Singh, C. J., Willson, T. M.; Collins, J. L.; Kliewer, S. A. *Proc. Nat. Acad. Sci.* **2000,** *97,* 7500-7502.
29. Balzer, I.; Hardeland, R. *Bot. Acta.* **1996,** *109,* 180-183
30. Kolar, J.; Machackova, I.; Eder, J.; Prinsen, E.; van Dongen, W.; van Onckelen, H.; Illnerova, H. *Phytochem.* **1997,** *8,*1407-1413.
31. Thorpe, T. A. *In vitro* Embryogenesis in Plants. Current Plant Science and Biotechnology in Agriculture; Kluwer Academic Publishers; Dordrecht, Netherlands, 1995.
32. Murch, S. J.; Choffe, K. L.; Victor, J. M. R.; Slimmon, T. Y.; KrishnaRaj, S.; Saxena, P. K. *Plant Cell Rep.* **2000,** *19,* 576-581.
33. Murch, S. J.; KrishnaRaj, S.; Saxena, P. K. *Plant Cell Rep.* **2000,** *19,* 698-704.

Chapter 14

Production of Vaccines and Therapeutics in Plants for Oral Delivery

L. M. Welter

AgriVax, Inc., 2250 Alcazar Street, Los Angeles, CA 90033

With in the last decade there has been enormous advancement in the utilization of transgenic plants or plant-derived products for mucosal delivery of vaccine and therapeutic proteins to the gastrointestinal tract. This review focuses on the unique aspects of mucosal immune system as it relates to mucosal delivery of vaccines or therapeutics using plants as edible dietary delivery systems. Edible plant delivery systems are attractive because they are safe, inexpensive, and easy to administer. However, in order to be efficacious they must survive the adverse environment of the stomach and be formulated to induce the appropriate immunological response of immunity or tolerance.

The mucosal immune system

The mucosal immune system (MIS) is the first line of defense against most pathogenic organisms where 95% of all infections occur. Secretory IgA (sIgA) has been the predominant immunoglobulin associated with protection of the gastrointestinal, respiratory and urogenital mucosal tissues against infections originating at these sites. Vaccinations for mucosal infections have been less successful than vaccinations for systemic infections presumably due to inadequate understanding of local mucosal immune responses. For current review of the field see Hayday and Viney (1) or Czerkinsky et al. (2) The two most studied mucosal inductive sites are found in the Peyer's patches of the intestinal track, often referred to as gut associated lymphoid tissue (GALT) and in the tonsillar/adenoid associated tissue often referred to as the nasal-associated lymphoid tissue (NALT). Both of these sites contain specialized cells (M-cells) responsible for uptake and transport of antigen as well as whole pathogenic organisms. M-cells initiate the first step of a mucosal immune response by transporting antigen across the epithelial barrier where it is processed and presented by antigen presenting cells (APC). It is believed that macrophages process foreign antigens and present them to a population of T lymphocytes called T helper cells (Th), which characteristically express the marker molecule CD4. Th1 cells mediate delayed-type hypersensitivity and are thought to assist T cells, which bear the molecular marker CD8 in becoming cytotoxic T lymphocytes (CTL), while Th2 cells collaborate with B lymphocytes in the production of antibodies. Th1 and Th2 cells can be distinguished from one another by the type of cytokines they produce. Specifically, Th1 cells secrete proinflammatory cytokines, such as interleukin-2 (IL2), γ-interferon (IFN-γ) and tumor necrosis factor (TNF) while Th2 cells produce anti-inflammatory cytokines, such as IL-4, IL-5, IL-6, IL-10 and IL-13. These cytokines mediate cellular interactions by being secreted from Th cells interacting with receptors on appropriate cells (i.e. CTLs express IL-2 receptors which enable them to respond to Th1 cells). The ratio of Th1 to Th2 cells generated during an immune response varies according to the pathogen.

Lymphocytes require two signals to become activated. One is antigen-specific and involves the recognition of degraded foreign antigen in association with major histocompatibility complex (MHC) class I or class II molecules by T-cell receptors (TCRs) (3). The second signal is not antigen-specific and is mediated by soluble cytokines. An important observation that lymphocytes receiving stimulation from specific antigens in the absence of the second signal, not only fail to become activated, but are refractory to further stimuli (4).

Initial stimulus received by the Th cell in association with specific signals, costimulatory molecules and specific cytokines determines the direction and progression of the immune response toward either a Th1 response (supporting a cell mediated response) or Th2 response (supporting a humoral response). Evidence supports both Th1 or Th2 cells or a combination of these cell types to induce antigen-specific sIgA responses. Transforming growth factor (TGF-β1) appears to be the most import cytokine signal for induction of B-cells to produce sIgA. The production of TGF-β1 subsequently results in a down regulation of IL-4 production which inhibits IgE production. TGF-β1 is also involved in another immunological occurrence referred to as tolerance.

One cannot develop an oral vaccine without taking into consideration the induction of tolerance. Tolerance is defined as a state of immunological hyporesponsiveness. The ability of T cells to specifically recognize foreign antigens while simultaneously being unresponsive or tolerant of self antigens is not totally understood, but appears to be determined by two selection processes which occur in the thymus. The failure of the mechanisms responsible for self tolerance results in autoimmune disease. The underlying causes for this failure are not understood although infections in genetically susceptible individuals may trigger disease *(5, 6)*. The effect of oral tolerance was first reported by Wells *(7)*, who observed that guinea pigs fed hen egg protein were resistant to anaphylaxis challenge with the same protein. The apparent mechanism of oral tolerance has been the subject of numerous reports. Tolerance appears to be associated with either antigen specific stimulation of suppressive Th cells in the GALT that secrete cytokines TGFß and IL-4 when antigen is administered at a low dose or clonal anergy and/or deletion when antigen is administered at a high dose *(8, 9, 10)*. Mucosal administration of antigens has the potential to induce oral tolerance or sensitization (priming) to the fed antigen. Factors that influence the type of response include nature of the antigen (particulate or soluble), susceptibility to degradation, dosage, absorption and prior exposure to the antigen. Induction of tolerance has been studied in a number of animal disease models used in the treatment of autoimmune diseases and allergens. Although these immunological events may not be mutually exclusive, when administering antigens to the mucosal sites it is import to take into consideration the desired immune response. For example, oral tolerance for the treatment of autoimmune disease could theoretically result in the priming and exacerbation of the disease.

Additional methods that have been explored to enhance the responses of mucosal immunization include experimental routes of administration, (i.e. nasal or transcutaneous), novel delivery systems (i.e. antigen encapsulation, DNA delivery, or liposome incorporation) and use of mucosal adjuvants such as saponin derivatives or bacterial toxins derived from *Vibrio cholera,* cholera toxin (CT) or derived from enterotoxigenic *E. coli* (ETEC), heat-labile enterotoxin (LT). None of these delivery methods compare with the simplicity cf administering a vaccine by ingesting an edible plant.

One obstacle that has made oral vaccination impractical has been the necessity of repeated administration of large doses of antigens required to establish a protective immune response. Utilization of plant-derived material circumvents this obstacle, because plants have the capacity to produce large quantities of the antigens. Thus the greatest potential markets for plant-derived proteins are those which require large quantities such as oral vaccines, antibodies and some therapeutics.

Although there are numerous reports of potential oral vaccine candidates, in reality there are only a few mucosal vaccines approved for use in humans. They include the oral poliovirus, oral *Salmonella typhi*, and oral rotavirus. Promising results obtained in advanced human clinical trials with a nasal vaccine against influenza could result in the first nasally administered vaccine for humans *(11,12)*. All of these mucosal vaccines are derived from live attenuated organisms, making their safety and stability a considerable issue. Oral delivery of vaccines administered by mixing with feed or water is the easiest method of mass vaccination of agricultural animals and has been incorporated into current agricultural production practices. Use of oral vaccines results not only in an economical savings of time and labor, but is also less stressful to the animal. Examples of oral vaccines for animals include a commercially available live attenuated Transmissible Gastroenteritis Virus (TGEV) vaccine for swine administered by mixing with ground corn, an avirulent Newcastle virus vaccine for poultry administered by mixing with cooked white rice, and an attenuated rabies vaccine for wild life administered by mixing with bait. Again the vaccines are derived from either live attenuated or genetically engineered organisms and are subject to the same safety concerns associated with any biologically derived vaccine.

Oral Vaccine Candidates Expressed in Plants

There are numerous publications of proteins expressed in plants for use as vaccines or therapeutics. The two main methods of production of proteins in plants include production of transgenic plants engineered to carry the gene of interest or production of plant viruses engineered to carry an epitope of the gene of interest. As with any recombinant derived vaccine the protective antigen or antigenic epitopes must be known. Both systems have their advantages and disadvantages. Some of these advantages and disadvantages are summarized in Table I.

Table I. Transgenic Plant and Plant Virus Expression Systems.

	Advantages	Disadvantages
Transgenic Plants	Low capital & production costs Unlimited supply No extraneous agents Easy storage & increased stability	Moderate development costs Long development period [b] Potential for transgene migration
Plant Viral Systems	Low development costs Unlimited supply No extraneous agents No transgene migration Short development period [a]	Size limitations Moderate production costs Potential viral instability

[a] Timeframe to produce recombinant plant virus as little as 1 month.

[b] Timeframe to produce transgenic plants from 3 months in potato to 18 months in corn.

[c] Large scale production requires plant infection & increased downstream processing

Proteins expressed using plant viral systems can be harvested from the infected plant tissues or small epitopes can be incorporated into the plant virus in which case the recombinant plant virus is harvested from the infected plant material. It is well documented that virus like particles (VLP) are highly immunogenic and are capable of inducing protective immunity (13). VLP are derived from self-assembling viral capsid proteins. Their particulate nature make them good subunit vaccine candidates as their particulate nature is better at stimulating an immune response than are soluble antigens. Vaccines expressed by recombinant plant virus particles (rPVP) may benefit from this immunological advantage, but this advantage must be weighed against the limited capacity of the system to carry complete antigenic genes. Very few vaccines are protective with a single antigenic epitope. Although transgenic plants have been developed that express multiple genes which result in functional proteins such as monoclonal antibodies or cholera toxin (CT), a single transgenic plant would not have to express all the antigenic genes for an edible vaccine. Rather the vaccine would be derived by blending individual transgenic plant lines expressing different antigens. The very nature of delivering the antigen in the plant tissues may actually facilitate its delivery by protecting it from degradation.

A summary of the production of viral, bacterial, or parasitic vaccine antigens expressed in plants is presented in Table II. Currently antigens from 15 viruses, 5 bacteria and 3 parasites have been expressed in plants with the list growing daily. The most common method of evaluation is immunogenicity in an animal model; however, pertinent evaluation needs to be conducted in a host animal challenge model (HACM) when available. Edible vaccines are extremely attractive for their ease of administration, storage, safety, and economical production. However, the crucial point to consider when evaluating an edible plant vaccine is its efficacy which must be equal to or better than currently available vaccines.

The first viral antigen expressed in a transgenic plant was the Hepatitis B surface antigen (HBsAg). Initial animal studies showed that the plant expressed antigen was immunogenic (14,15). An edible plant vaccine consisting of HBsAg expressed in lettuce has recently been shown to be immunogenic when fed to humans (16). The HACM for determining the efficacy of an edible HBsAg vaccine is very costly as efficacy tests must be conducted in primates, as the host animal is man. However, these findings are notable as they concur with previous findings of immunogenicity of edible plant vaccines for Norwalk virus or E. coli LTB in humans (17,18). Of all the viruses listed in Table II five demonstrated efficacy in an animal challenge model, but only two out of the five were compared to currently available vaccines and only three were fed as an edible plant vaccine.

Our laboratory was the first to demonstrate the feasibility of an edible vaccine in a host animal challenge model using transgenic potato plants for transmissible gastroenteritis virus (TGEV) for swine (19). A TGEV truncated

Table II. Vaccine Antigens Expressed in Plants.

Pathogen	Gene	Plant System [a]	Evaluation [b]	Ref
Viral				
BVDV	E2	Barley	Exp	20*
CPV	VP2	PPV	Im-M:ip-F: nAb(+); Im-Rb:im-F: nAb(+)	21
FMDV	VP1	CPMV	Exp	22
	VP1	Arabidopsis	ACD-M:ip-F:P	23
	VP1	Alfalfa	ACD-M:ip-F,o:P	24
HBV	HBsAg	Tobacco	Exp	14
	HBsAg	Tobacco	Im-M:ip-F:Ab(+):TPR(+)	15
	HBsAg	Potato	Exp	25
	HBsAg	Lettuce; Lupin	Im-H:o:Ab(+)	16
			Im-M:o:Ab(+)	16
hCMV	gB	Tobacco	Exp	26
HIV	gp120	AlMV	Im-M:ip-F:Ab(+)	27
	gp41	CPMV	Im-M:sq-Al:nAb(+)	28
	gp41	CPMV	Exp	29
	gp120	TBSV	Im-M:sq-I:Ab(+)	30
	gp41	CPMV	Im-M:n-CT,o-CT:Ab(+)	31
	p24	TBSV	Exp	32*
	gp41	CPMV	Im-M:sq- Al,F,R,QA,AP:Ab(+):TPR(+)	33
HRV-14	VP-1	CPMV	Im-Rb:im-F,sq-F: Ab(+)	29; 34
MEV	VP2	CPMV	HACD-Mk:sq-QA & Al:P	35
MHV	S	TMV	HACD-M:n,sq-R:P	36
NV	NVCP	Tobacco; Potato	Im-M:o-CT:Ab(+)	13
	NVCP	Potato	Im-H:o:Ab(+)	17
PRRS	ORF5[s]	Tobacco	Exp	37*

Rabies	G	Tomato	Exp	38
	G	AIMV	ACD-M:ip,o:P	39
RHDV	VP60	Potato	HACD-Rb:sq,im:P	40
RSV	F	Tomato	Im-M:o:Ab(+)	41
TGEV	S	Potato	HACD-S:o-mLT:P	19
	S	Arabidopsis	Im-M:ip:nAb(+)	42
	S	Alfalfa; Tobacco	Im-S:ip,o:Ab(+)	43*
	S	Alfalfa; Tobacco	Im-M:im:Ab(+)	44*
	S^s	Alfalfa; Tobacco	Exp	45*
	S	Tobacco	Im-S:ip-MO:Ab(+)	46
	S^s	Corn	HACD-S:o:P	47
Bacterial				
V. cholera	CTA:CTB	Tobacco	Exp	48
	CTB	Potato	Exp	49
	CTB	Potato	ACD-M:o:P	50
	CTB	Potato	ACD-M:o:P	51
E. coli	LTB	Potato	Im-M:o:Ab(+)	52
	LTB^s	Potato	Im-H:o:Ab(+)	18
	LTB^s	Potato	ACD-M:o:P	53
P. aeruginosa	OMF	CPMV	Im-M:sq-QA,QS-21,F:Ab(+)	54
	OM	CPMV;TMV	ACD-M:sq:P	55
S. aureus	FnBP	CPMV;PVX	Im -M:sq-QS-21:Ab(+)	56
			Im-R:sq-F:Ab(+)	56
	FnBP	CPMV	Im-M:n-ISCOM,o-Lip-QS-21:Ab(+)	57
S. mutans	SpaA	Tobacco	Exp	58

Continued on next page.

Table II. Vaccine Antigens Expressed in Plants.

Parasite				
C. parvum	P23	Potato	Exp	59
	CP15/60	Potato	Exp	
I. suis	SAP	Potato	Exp	60
Malaria	CSP	TMV	Exp	61

[a] All plant systems are either transgenic plants or the indicated plant virus expressing an epitope or portion of the gene of interest.

[b] Exp, analyzed for protein expression and characterization; Im, immunogenicity tested in an animal model; nAb, neutralizing antibodies analysis; Ab, antibody response analysis; TPR, T cell proliferative response analysis.

Animal species are abbreviated as M, mouse; Mk, mink; H, human; R, rat; Rb, rabbit; S, swine.

Route of administration is abbreviated as im, intramuscular; ip, intraperitoneal; n, nasal; o, oral; sq, subcutaneous.

Adjuvant abbreviation AP, AdjuPrime (Pierce); Al, aluminum hydroxide; CT, *Vibrio* cholera toxin; F, Freund's adjuvant; I, ISCOM a saponin adjuvant;); mLT, mutant *E. coli* heat labile enterotoxin LT(R192G); MO, mineral oil; QA, Quil A, a saponin derived adjuvant; QS-21 a saponin derived adjuvant; R, Ribi adjuvant system (MPL+TDM).

Animal results abbreviated as ACD, animal challenge data; HACD, host animal challenge data; P, protection observed in animal challenge model; P(-), no protection observed in animal challenge model.

[s] Represents synthesized S gene to contain plant friendly codon usage.

* Unpublished results presented at the International Molecular Farming Conference August 29 to September 1 1999, London, Ontario, Canada.

amino-terminal fragment of the S protein was expressed in transgenic potato plants and fed three doses as a vaccine, either with or without a mucosal adjuvant (*E. coli* LT (R192G)). Pigs, which received both the TGEV transgenic potato vaccine and the mucosal adjuvant, and subsequently challenged with virulent TGEV, exhibited reduced morbidity (46%), mortality (40%) and TGEV isolation (60%) and increased average daily weight gains and TGEV serum neutralization titers. Animals receiving only TGEV transgenic plant material as a vaccine were not protected, however the amount of S protein expression in the plants was very low. Our initial attempts to express the full-length S gene failed, as the gene is very large and was not optimized for plant codon usage. Others have since demonstrated expression or immunogenicity of the TGEV S protein expressed in arabidopsis, alfalfa, or tobacco *(42,43,44,45,46)*. In these examples re-synthesizing the S gene to contain plant friendly codon usage resulted in increased expression. Recently, a re-synthesized S gene was expressed in corn and 50 g of ground corn material containing approximately 1mg of S protein was feed to pigs for 10 days. Animals were challenged with virulent TGEV two days later. The pigs receiving the corn expressed S protein were protected against challenge. The protection observed was comparable to that of a commercially available modified live vaccine *(47)*. Testing in a host animal challenge model and comparison to the currently licensed oral vaccine offers the most conclusive support of efficacy of edible plant vaccines.

An edible plant vaccine for foot and mouth disease virus (FMDV) has also demonstrated efficacy in an animal challenge model *(24)*. Transgenic alfalfa plants expressing the VP1 protein of FMDV was either feed or administered intraperitoneally to mice which were subsequently challenged with FMDV and monitored for absence of viremia as an indicator for protection. Mice immunized intraperitoneally exhibited a protection rate of 77% to 80%, whereas mice feed the FMDV transgenic plant material exhibited a protection rate of 66% to 75%. Mice are not a target species for FMDV, but evaluation of the vaccine in a HACM is difficult as it is highly contagious and there are strict regulatory and containment requirements for FMDV. This evidence supports the use of edible plant vaccines for yet another agriculturally important animal disease.

Efficacy of three additional plant expressed vaccines has been demonstrated in host animal challenge models using mice and mink. All were antigenic epitopes expressed by rPVP. In a HACM for mouse hepatitis virus (MHV), mice were immunized by intranasal or subcutaneous route with tobacco mosaic virus (TMV) expressing an epitope of the MHV S gene *(36)*. Animals receiving the rPVP vaccine were protected against lethal challenge. In another HACM the efficacy of a cowpea mosaic virus (CPMV) expressing an epitope of the VP2 capsid protein of mink enteritis virus was compared to a commercially available vaccine *(35)*. The commercially available vaccine is composed of inactivated virus derived from cell culture, adjuvanted with aluminum hydroxide gel, and is protective as a single subcutaneous dose. Animals vaccinated with a single subcutaneous dose containing 1 mg of purified rPVP adjuvanted with Quil A and aluminum hydroxide gel were protected from clinical disease, but still shed virus as compared to the commercially vaccinated animals that were protected from clinical disease and did not shed any virus post challenge. Both of these sets of experiments demonstrate the efficacy of plant-derived vaccines, but fail to capitalize on the most attractive feature of plant-derived vaccines which is edible delivery. In another HACM another plant virus system utilizing the alfalfa mosaic virus (AIMV) or TMV expressing epitopes of the rabies virus glycoprotein was evaluated in mice (39). The researchers evaluated rPVP vaccine administered by either intraperitoneal route, oral gastric intubation of purified rPVP, or oral feeding of plant leaves containing rPVP. Animals receiving the rPVP by intraperitoneal route were partially protected against lethal challenge as exhibited by a delayed onset of disease and increased survival rate of 40% as compared to control animals which developed clinical signs of the disease earlier with none surviving. Animals receiving 250 μg of rPVP by oral gastric intubation exhibited increased rabies specific serum IgG and IgA responses. However, these animals exhibited lower titer of rabies specific fecal IgA as compared to animals fed plant material containing an estimated dose of 25 μg of rPVP. The researchers did not present data comparing the serum antibody levels of animals fed the plant leaves containing rPVP. Both orally fed groups were subsequently challenged intranasally with an attenuated rabies virus in an animal challenge model which monitors average daily weight gains. Both orally vaccinated groups showed comparable average daily weight gains post challenge. However, these results cannot be compared with those obtained in the lethal animal challenge model used for the intraperitoneal vaccinated group.

Four points to consider in the development of a plant derived rabies vaccine are; 1) the normal route of rabies transmission is not by the mucosal route, but by infection via a bite wound, hence a systemic response as opposed to sIgA may be a primary host defense 2) standard potency tests are defined by the National Institutes of Health and by the European Pharmacopoeia with international standard vaccine available from the WHO, 3) the standard established for measuring rabies vaccine duration of immunity is defined by rabies specific serum neutralizing antibody titers and challenge with rabies virus, generally the percentage of sero-conversion and the mean level of antibody allow a good prognosis for survival to challenge, and 4) an oral rabies vaccine has been successfully used to orally vaccinate wildlife by incorporating an attenuated rabies virus into bait (62, 63,64). Thus the rabies disease model offers international standardized in vivo and in vitro tests for determining efficacy and an oral rabies vaccine has demonstrated efficacy. These tools offer a great opportunity to evaluate the efficacy of a plant derived rabies vaccine.

The first bacterial antigen expressed in plant was the surface protein antigen (SpaA) of Streptococcus mutans in tobacco. SpaA fed to mice induced a mucosal response that reacted with S. mutans (58). Two highly homologous bacterial exotoxins for ETEC or V. cholera expressed in transgenic plants have been evaluated in animal challenge models. In both cases mice were fed transgenic plant material expressing either a codon optimized version of LTB derived from *E. coli (53)* or CTB derived from *V. cholera (50)*. In both cases the animals were challenged with the homologous exotoxin and protected as defined as a reduction in fluid accumulation in the gut or small intestine. In both studies animals fed the transgenic plant vaccine developed specific and protective mucosal antibody responses. Human clinical trials have been completed in which 50 or 100 g of LTB transgenic potatoes representing a dose of 0.4 to 1.1 mg of LTB were fed to human volunteers on days 0, 7, and 21. Ten out of 11 volunteers ingesting the transgenic LTB vaccine developed a 4-fold increase in anti-LTB IgG titers and 6 out of the 11 developed a 4-fold increase in anti-LTB IgA titers. None of the volunteers ingesting wild-type potatoes developed significant increase of anti-LTB titers. This represents the first demonstration of immunogenicity of an edible plant vaccine in humans *(18)*.

Finally, plant expressed vaccines containing antigens from parasites have been developed. In one example an epitope for the circumsporozoite protein for malaria was expressed using the TMV system *(61)*. Our laboratory has expressed in transgenic potatoes coccida antigens for either *Isospora suis*, a causative agent of diarrhea in swine, or *Cryptosporidium parvum*, a causative agent of diarrhea in humans and cattle (unpublished results). An oral vaccine containing the *I. suis* sporozoite attachment factor (SAP) has been shown to reduce disease in a HACM *(60)*. Neutralizing serum antibodies recognizes SAP expressed by transgenic potato plants. *C. parvum* antigens for CP15/60 or p23 have also been expressed in transgenic potatoes. These antigens are associated with parasite motility and are thought to play a role during infection. However, preliminary evaluation of these antigens as oral vaccine candidates did not show protection in a HACM. There are no approved vaccines for these parasites.

Therapeutic Candidates Expressed in Plants

There are numerous examples demonstrating the expression of therapeutic proteins in transgenic plants. A list of some of these proteins is presented in Table III. Areas in which plant expression systems offer tremendous opportunities are in the production of monoclonal antibodies (MAb) and toleragens. Cost prohibitive expense of production of MAbs has precluded their use in some therapeutic applications. Production in plants significantly decreases the cost and thus opens the opportunity for use of MAbs in topical therapeutics. Currently there are a number of MAbs expressed in plants being developed for the prevention of a dental caries, the prevention of sexually transmitted diseases and the prevention of conception. The first MAb expressed in plants for oral delivery in humans was sIgA against the surface adhesion molecule for *Streptococcus mutans* responsible for bacterial colonization and subsequent development of dental caries in humans *(65)*. Expression of this MAb represented a momentous feat as individual transgenic plants representing the various antibody components for the kappa chain, immunoglobulin A-G heavy chain, a joining chain, and secretory component were developed. Through successive sexual crosses between these transgenic plants researchers were able to isolate plants that expressed all four proteins simultaneously and also demonstrated the ability of the plant expressed proteins to assemble into functional secretory immunoglobulin that recognized SA I/II *(65)*. Topical application of this plant expressed anti-SAI/II sIgA was tested in humans and demonstrated specific protection against oral streptococcal colonization *(66)*. This is the first demonstration of a topical passive immunotherapeutic expressed in plants. The amount used in this clinical trial consisted of six applications of 22.5 mg of plantibody per treatment. The application was well tolerated with no adverse reactions or reactivity (i.e. anti-plant antibodies) observed.

Additional antibodies expressed in plants are being developed for the prevention of sexually transmitted diseases and prevention of conception. An IgG MAb directed to the glycoprotein B of the herpes simplex virus 2 (HSV-2) has been expressed in soybeans. Plant expressed anti-HSV2 MAb maintained comparable biological characteristics of stability and serum neutralization as their mammalian expressed counterpart. Topical application of the plant expressed anti-HSV2 MAb was able to prevent HSV infection in a mouse model as determined by the presence or absence of viral lesions and the isolation of virus shedding *(81)*. At a 10 µg dose both the plant and mammalian expressed anti-HSV2 MAbs were able to prevent virus shedding. The plant derived anti-HSV2 MAb was able to prevent HSV lesions at a 1 µg dose as compared to the mammalian expressed anti-HSV2 MAb which required a 10 µg dose to prevent HSV lesions in all animal. Antibodies directed against sperm are currently being expressed in corn *(82)*. Combinations of plant expressed antibodies directed against sexually transmissible disease agents and sperm are being developed as a topical gel for use in the prevention of disease and conception.

Other examples in which plant derived therapeutics have demonstrated efficacy in animal models is in the prevention of autoimmune diseases by induction of oral tolerance. For current review of the field see proceedings of the conference "Oral Tolerance: Mechanisms and Applications" *(86)*. Induction of tolerance has been studied in a number of animal disease models for use in the treatment of autoimmune diseases, such as: multiple sclerosis by tolerizing with myelin basic protein (MBP) or proteolipid protein (PLP); rheumatoid arthritis by tolerizing with type II collagen; uveoretinitis by tolerizing with S–antigen or interphotoreceptor retinoid–binding protein; type I diabetes by tolerizing with insulin (INS) or glutamate decarboxylase (GAD); myasthenia gravis by tolerizing with acetylcholine receptor; and thyroiditis by tolerizing with thyroglobulin. Induction of tolerance is also currently being investigated for prevention of transplant rejection by tolerizing with alloantigen or MHC peptide, and for prevention of allergic immune responses including allergic reactions to cats and bee stings by tolerizing with Fel d1 or PLA–2 peptides, respectively *(87,88)*. Induction of tolerance to autoantigens has recently been applied in human clinical trials for multiple sclerosis, rheumatoid arthritis, uveoretinitis, and type I diabetes *(87)*. Initial results of phase I/II studies showed that a small portion of patients responded favorably to the treatment that consisted of feeding bovine brain derived MBP or bovine derived type II collagen for the treatment of multiple sclerosis or rheumatoid arthritis, respectively *(89,90)*. Initial results supported the use of oral tolerance for the treatment of autoimmune diseases. This type of treatment is appealing because of the lack of toxicity and low incidence of adverse side effects observed. Results of large scale phase III trials are in progress, but so far there has been no observed toxicity or exacerbation of the disease.

Demonstration of clinical efficacy in these trials may be limited by practical problems, in that large quantities of antigens (i.e. mg to kg) are required to induce oral tolerance in experimental animals as well as humans by the oral route. Thus plants are extremely attractive as a delivery vehicle of autoantigens.

Tobacco or potato plants expressing the autoantigen GAD given as a dietary supplement (calculated at approximately 1 to 1.5 mg of GAD/dose) inhibited the development of diabetes in non-obese diabetic (NOD) mice, an animal model for diabetes *(83,84)*. It has also been shown that the feeding of plants expressing a GAD67 or INS-CTB fusion proteins were also able to inhibit the development of diabetes in the NOD mouse model *(51)*. It has previously been demonstrated that conjugation or fusion of antigen to the CTB subunit reduces the required amount of antigen for oral tolerization *(91)*. The actual mechanism is unknown, but it is speculated that the CTB acts as a carrier molecule for the attached antigen. However, the long term clinical effectiveness of administering a CTB fusion antigen is unknown as CTB is highly immunogenic. In fact NOD mice fed CTB-INS or CTB-GAD transgenic potatoes developed serum and intestinal antibody responses to CTB and elevated anti-INS and anti-GAD serum antibody responses as compared to those animals fed GAD, INS or wild-type potatoes. Animals were fed transgenic plant material containing either 30 μg of INS, 20μg INS-CTB, 3 μg GAD, 2 μg GAD-CTB or wild-type potato. Only the INS-CTB and GAD-CTB fed animals showed a reduction in their insulitis score. These results conflict with a previous studies which demonstrated that feeding of plant expressed GAD material inhibited diabetes in the same NOD mouse model. An obvious difference is the amount of antigen fed. A high dose of 1 to 1.5 mg of GAD/dose was shown to be an effective treatment, whereas animals fed a low dose containing either 3 μg of GAD/dose or 30 μg of INS/dose was not effective. Simultaneous feeding of INS-CTB and GAD-CTB plant material resulted in a synergistic effect and substantial reduction in insulitis scores as compared to feeding either antigen alone. These experiments support the use of transgenic plants for the treatment of autoimmune disease, however high expression levels must be obtained. Other tolerogenic antigens that have been expressed in transgenic plants include the autoantigens MHC class II and hMBP and the allergen for house dust mite *(84,85)*.

Table III. Therapeutic Antigens Expressed in Plants.

Therapeutic	Gene	Plant System[a]	Evaluation[b]	Ref
Cancer vaccine	scFv	TMV	ACD-M:sq:P	67
hIL-10	Same	Tobacco	Exp	68
	Same	Tobacco	Exp	69*
	Same	Tobacco	Exp	70*
hGrowth Hormone	Same	Tobacco	Exp	71*
hLactoferrin	Same	Rice; Tobacco Potato; Tomato	Exp	72*; 73;74*
hLysozyme	Same	Rice	Exp	74*
hLactoferricin	Same	Rice	Exp	74*
α-lactalbumin	Same	Tobacco	Exp	75
β-casein	Same	Potato	Exp	76
hAlpha-1-antitrypsin	Same	Rice	Exp	77
hserum Albumin	Same	Potato; Tobacco	Exp	78
Zona pellucida	ZP3 pep	TMV	ACD-M:ip-R:P(-)	79
Monoclonal Antibodies				
MAb-CEA	α-CEA	Rice	Exp	80*
MAb-HSV	α-gB	Soybean	ACD-M:t:P	81
MAb-Sperm	α-Sperm	Corn	Exp	82*
MAb-S mutans	α-(SA)CSA	Tobacco	Exp	65
		Tobacco	HACD-H:t:P	66
Autoantigens/Allergens				
GAD67	mGAD	Tobacco; Potato	ACD:M:o:P	83; 84
MHC	mMHC	Tobacco	Exp	84
hInsulin-CTB fusion	CTB-hINS	Potato	ACD-M:o:P	51

Continued on next page.

hGAD-CTB fusion	CTB-hGAD	Potato	ACD-M:o:P	51
hMBP	hMBP	Potato	Exp	85
House Dust Mite Ag	DerP1	Potato	Exp	85

[a] All plant systems are either transgenic plants or the indicated plant virus expressing an epitope or portion of the gene of interest.

[b] Exp, analyzed for protein expression and characterization.

Animal species are abbreviated as M, mouse; H, human

Route of administration is abbreviated as ip, intraperitoneal; o, oral; t, topical

Animal results abbreviated as ACD, animal challenge data; HACD, host animal challenge data; P, protection observed in animal challenge model; P(-), no protection observed in animal challenge model.

Future Directions

Currently there are no approved vaccines or therapeutics expressed in plants for use in animals or humans. On April 5-6,2000 the U. S. government regulatory agencies for human (CBER/FDA) and animal (USDA APHIS) health held a Public Hearing on Plant-Derived Biologics to address regulatory and policy issues related to the manufacture, distribution, and use of biological products derived from plants *(92)*. This recognition of the plant-derived vaccines, therapeutics and diagnostics as viable commercial products is an important step in transforming this new technology from a concept to a reality. There are currently over 12 companies that are currently developing commercial products expressed in plants. In 1999 four of these companies were in field tests for plant-derived biologics (Large Scale Biology, ProdiGene, Applied Phytologics, and Monsanto). One of these companies, ProdiGene, anticipates the first license of a plant derived animal vaccine within the next two to three years *(93)*.

Proof of concept has been demonstrated for numerous plant expression systems including the first commercially available product, avidin expressed in corn, (Sigma Chemical Company). Unfortunately the first commercially available vaccine or therapeutic expressed in plants is still a few years away.

Issues that will need to be addressed include: increased levels of expression either by tissue specific or inducible promoters; increased stability and accumulation of foreign proteins in plants; variable levels of protein expression due to variations during growth conditions; evaluation of potential adverse side affects due to differences in plant glycosylation (i.e. allergies); the potential for development of tolerance with orally delivered subunit vaccines; intellectual property issues that may preclude entry into commercial markets; minimizing processing cost to minimize production costs; establishment of regulatory policies and finally gaining public acceptance of plant derived therapeutics and vaccines. As these issues are addressed the number of vaccines and therapeutics expressed in plants will continue to grow.

References

1   Hayday A, Viney JL. Science. 2000 Oct 6;290(5489):97-100.
2   Czerkinsky C, Anjuere F, McGhee JR, George-Chandy A, Holmgren J, Kieny MP, Fujiyashi K, Mestecky JF, Pierrefite-Carle V, Rask C, Sun JB. Immunol Rev. 1999 Aug;170:197-222.
3   Germain RN. Cell. 1994 Jan 28;76(2):287-99.
4   Mueller DL, Jenkins MK, Schwartz RH. Annu Rev Immunol. 1989;7:445-80

187

5   Bernard CC, Kerlero de Rosbo N. Curr Opin Immunol. 1992 Dec;4(6):760-5.
6   Sprent J, Gao EK, Webb SR Science. 1990 Jun 15;248(4961):1357-63.
7   Wells,H.G. J. Infect. Dis. 1911, 9:147-171.
8   Critchfield JM, Racke MK, Zuniga-Pflucker JC, Cannella B, Raine CS, Goverman J, Lenardo MJ. Science. 1994 Feb 25;263(5150):1139-43.
9   Chen Y, Inobe J, Kuchroo VK, Baron JL, Janeway CA Jr, Weiner HL. Proc Natl Acad Sci U S A. 1996 Jan 9;93(1):388-91.
10   Gienapp I, Cox K, Javed N, Whitacre C. Ann N Y Acad Sci. 1996 Feb 13;778:382-3.
11   Belshe RB, Gruber WC, Mendelman PM, Cho I, Reisinger K, Block SL, Wittes J, Iacuzio D, Piedra P, Treanor J, King J, Kotloff K, Bernstein DI, Hayden FG, Zangwill K, Yan L, Wolff M. J Pediatr. 2000 Feb;136(2):168-75.
12   Belshe RB, Gruber WC, Mendelman PM, Mehta HB, Mahmood K, Reisinger K, Treanor J, Zangwill K, Hayden FG, Bernstein DI, Kotloff K, King J, Piedra PA, Block SL, Yan L, Wolff M. J Infect Dis. 2000 Mar;181(3):1133-7.
13   Mason HS, Ball JM, Shi JJ, Jiang X, Estes MK, Arntzen CJ. Proc Natl Acad Sci U S A. 1996 May 28;93(11):5335-40.
14   Mason HS, Lam DM, Arntzen CJ. Proc Natl Acad Sci U S A. 1992 Dec 15;89(24):11745-9.
15   Thanavala Y, Yang YF, Lyons P, Mason HS, Arntzen C. Proc Natl Acad Sci U S A. 1995 Apr 11;92(8):3358-61.
16   Kapusta J, Modelska A, Figlerowicz M, Pniewski T, Letellier M, Lisowa O, Yusibov V, Koprowski H, Plucienniczak A, Legocki AB. FASEB J. 1999 Oct;13(13):1796-9.
17   Tacket CO, Mason HS, Losonsky G, Estes MK, Levine MM, Arntzen CJ. J Infect Dis. 2000 Jul;182(1):302-5.
18   Tacket CO, Mason HS, Losonsky G, Clements JD, Levine MM, Arntzen CJ. Nat Med. 1998 May;4(5):607-9.
19   Welter, L.M., Mason H.S., Lu, W., Lam, D.M-K., and Welter, M. (1996) Effective Immunization of Piglets with Transgenic Potato Plants Expressing a Truncated TGEV S Protein. CHI Symposium on Vaccines: New Technologies and Applications March 1996. McLean, Virginia..
20   F. Eudes*1, S. A. Gilbert2, S. Acharya1 and A. Laroche1. 1Department of Crop Science, Lethbridge Research Centre, Lethbridge, Alberta T1J 4B1; 2Animal Diseases Research Institute, Lethbridge, Alberta T1J 3Z4.(unpublished results).
21   Fernandez-Fernandez MR, Martinez-Torrecuadrada JL, Casal JI, Garcia JA. FEBS Lett. 1998 May 8;427(2):229-35.
22   Usha R, Rohll JB, Spall VE, Shanks M, Maule AJ, Johnson JE, Lomonossoff GP. Virology. 1993 Nov;197(1):366-74.
23   Carrillo C, Wigdorovitz A, Oliveros JC, Zamorano PI, Sadir AM, Gomez N, Salinas J, Escribano JM, Borca MV. J Virol. 1998 Feb;72(2):1688-90.
24   Wigdorovitz A, Carrillo C, Dus Santos MJ, Trono K, Peralta A, Gomez MC, Rios RD, Franzone PM, Sadir AM, Escribano JM, Borca MV. Virology. 1999 Mar 15;255(2):347-53.

25 Ehsani P, Khabiri A, Domansky NN. Gene. 1997 Apr 29;190(1):107-11.

26 Tackaberry ES, Dudani AK, Prior F, Tocchi M, Sardana R, Altosaar I, Ganz PR. Vaccine. 1999 Aug 6;17(23-24):3020-9.

27 Yusibov V, Modelska A, Steplewski K, Agadjanyan M, Weiner D, Hooper DC, Koprowski H. Proc Natl Acad Sci U S A. 1997 May 27;94(11):5784-8.

28 McLain L, Durrani Z, Wisniewski LA, Porta C, Lomonossoff GP, Dimmock NJ. Vaccine. 1996 Jun;14(8):799-810.

29 Porta C, Spall VE, Lin T, Johnson JE, Lomonossoff GP. Intervirology. 1996;39(1-2):79-84.

30 Joelson T, Akerblom L, Oxelfelt P, Strandberg B, Tomenius K, Morris TJ. J Gen Virol. 1997 Jun;78 (Pt 6):1213-7.

31 Durrani Z, McInerney TL, McLain L, Jones T, Bellaby T, Brennan FR, Dimmock NJ. J Immunol Methods. 1998 Nov 1;220(1-2):93-103.

32 Guichang, Z.*, Rodrigues, L.[2]; Murdin, L.[2]; Leung,C.2; Rovinski, B[2];White, K. A.[1] Department of Biology, York University, Toronto, Ontario, Canada; [2]Pasteur Merieux Connaught Canada, Toronto, Ontario, Canada (unpublished results).

33 McInerney TL, Brennan FR, Jones TD, Dimmock NJ. Vaccine. 1999 Mar 17;17(11-12):1359-68.

34 Porta C, Spall VE, Loveland J, Johnson JE, Barker PJ, Lomonossoff GP. Virology. 1994 Aug 1;202(2):949-55.

35 Dalsgaard K, Uttenthal A, Jones TD, Xu F, Merryweather A, Hamilton WD, Langeveld JP, Boshuizen RS, Kamstrup S, Lomonossoff GP, Porta C, Vela C, Casal JI, Meloen RH, Rodgers PB. Nat Biotechnol. 1997 Mar;15(3):248-52.

36 Koo M, Bendahmane M, Lettieri GA, Paoletti AD, Lane TE, Fitchen JH, Buchmeier MJ, Beachy RN. Proc Natl Acad Sci U S A. 1999 Jul 6;96(14):7774-9.

37 Rymerson[1], R.; Zhang[2], J.; Yoo[3], D.;Erickson[2] L.; Brandle[1], J. [1]Agriculture and Agrifood Canada, ON, Canada; [2]Department of Plant Agriculture, University of Guelph, Guelph ON, Canada; [3]Department of Pathobiology, University of Guelph, Guelph ON, Canada (unpublished results).

38 McGarvey PB, Hammond J, Dienelt MM, Hooper DC, Fu ZF, Dietzschold B, Koprowski H, Michaels FH. Biotechnology (N Y). 1995 Dec;13(13):1484-7.

39 Modelska A, Dietzschold B, Sleysh N, Fu ZF, Steplewski K, Hooper DC, Koprowski H, Yusibov V. Proc Natl Acad Sci U S A. 1998 Mar 3;95(5):2481-5.

40 Castanon S, Marin MS, Martin-Alonso JM, Boga JA, Casais R, Humara JM, Ordas RJ, Parra F. J Virol. 1999 May;73(5):4452-5.

41 Sandhu JS, Krasnyanski SF, Domier LL, Korban SS, Osadjan MD, Buetow DE. Transgenic Res. 2000 Apr;9(2):127-35.

42 Gomez N, Carrillo C, Salinas J, Parra F, Borca MV, Escribano JM. Virology. 1998 Sep 30;249(2):352-8.

43 Erickson, L.; Nagy1, E.; Tuboly[1], T.; Yu, W.; Bailey, A.; Du, S. Department of Plant Agriculture, Ontario Agricultural College, and Department of

Pathobiology[1], Ontario Veterinary College, University of Guelph, Guelph, Ontario (unpublished results).

44 Bailey, A.; Yu,W ; Tuboly, T; Nagy, E.;Erickson, L.. Plant Agriculture Department - Biotechnology Division, University of Guelph, Guelph, Ontario, Canada (unpublished results).

45 Yu, W.J.; Tuboly[2] T.;Bailey[2],A.;Du[2],S. ;Nagy[1],E. ;Erickson[2],L. Department of Plant Agriculture[2], Ontario Agricultural College, and Department of Pathobiology[1], Ontario Veterinary College, University of Guelph, Guelph, Ontario, Canada (unpublished results).

46 Tuboly T, Yu W, Bailey A, Degrandis S, Du S, Erickson L, Nagy E. Vaccine. 2000 Apr 3;18(19):2023-8.

47 Jilka, J.; ProdiGene, College Station Texas Unpublished results presented at Plant-Derived Biologics Seminar and Public Hearing on Plant-Derived Biologics, April 5 & 6, 2000, Ames, Iowa USA. (unpublished results).

48 Hein MB, Yeo TC, Wang F, Sturtevant A. Ann N Y Acad Sci. 1996 May 25;792:50-6.

49 Arakawa T, Chong DK, Merritt JL, Langridge WH. Transgenic Res. 1997 Nov;6(6):403-13.

50 Arakawa T, Chong DK, Langridge WH. Nat Biotechnol. 1998 Mar;16(3):292-7.

51 Arakawa T, Yu J, Langridge WH. Adv Exp Med Biol. 1999;464:161-78.

52 Haq TA, Mason HS, Clements JD, Arntzen CJ. Science. 1995 May 5;268(5211):714-6.

53 Mason HS, Haq TA, Clements JD, Arntzen CJ. Vaccine. 1998 Aug;16(13):1336-43.

54 Brennan FR, Jones TD, Gilleland LB, Bellaby T, Xu F, North PC, Thompson A, Staczek J, Lin T, Johnson JE, Hamilton WD, Gilleland HE Jr. Microbiology. 1999 Jan;145 (Pt 1):211-20.

55 Gilleland HE, Gilleland LB, Staczek J, Harty RN, Garcia-Sastre A, Palese P, Brennan FR, Hamilton WD, Bendahmane M, Beachy RN. FEMS Immunol Med Microbiol. 2000 Apr;27(4):291-7.

56 Brennan FR, Jones TD, Longstaff M, Chapman S, Bellaby T, Smith H, Xu F, Hamilton WD, Flock JI. Vaccine. 1999 Apr 9;17(15-16):1846-57.

57 Brennan FR, Bellaby T, Helliwell SM, Jones TD, Kamstrup S, Dalsgaard K, Flock JI, Hamilton WD. J Virol. 1999 Feb;73(2):930-8.

58 Curtiss, R.I.; Cardineau, G.A.World Patent Application 1990, WO 90/02484.

59 Welter, L.; Kuss, R.;Bischoff, D. AgriVax, Inc. Los Angeles, CA. unpublished results).

60 Quick D.P., Welter M.W., Welter C.J., Welter L.M. and Steger A.M. ISOSPORA SUIS VACCINE. US patent, 5,861,160 January 1999.

61 Turpen TH, Reinl SJ, Charoenvit Y, Hoffman SL, Fallarme V, Grill LK. Biotechnology (N Y). 1995 Jan;13(1):53-7.

62 Manual of standards for diagnostic tests & vaccines, 3rd Ed., 1996, Office International des Epizooties, Paris, France

63 Mackowiak M, Maki J, Motes-Kreimeyer L, Harbin T, Van Kampen K Adv Vet Med. 1999;41:571-83.

64 Bruyere V, Vuillaume P, Cliquet F, Aubert M. Vet Res. 2000 May-Jun;31(3):339-45.

65 Ma JK, Hiatt A, Hein M, Vine ND, Wang F, Stabila P, van Dolleweerd C, Mostov K, Lehner T. Science. 1995 May 5;268(5211):716-9.

66 Ma JK, Hikmat BY, Wycoff K, Vine ND, Chargelegue D, Yu L, Hein MB, Lehner T. Nat Med. 1998 May;4(5):601-6.

67 McCormick AA, Kumagai MH, Hanley K, Turpen TH, Hakim I, Grill LK, Tuse D, Levy S, Levy R. Proc Natl Acad Sci U S A. 1999 Jan 19;96(2):703-8.

68 Spurgeon D. BMJ. 1999 Jul 17;319(7203):143.

69 Menassa[1], R.; Jevnikar[2], A.; Ma[2], S.; Brandle[1], J. [1] Southern Crop Protection and Food Research Center, Agriculture and AgriFood Canada, London, Ontario CANADA, [2]MOTS- Transplantation and Immunobiology Group London Health Sciences Center, London Ontario, Canada (unpublished results).

70 Menassa1, R.; Armstrong[1],J.; Nguyen[1], V.; Jevnikar[3],A.;Miki[2],B.; Brandle[1],J. Agriculture and Agri-Food Canada; [1]Southern Crop Protection and Food Research Centre, London, Ontario, [2]ECORC,Ottawa, Ontario, [3]London Health Sciences Centre, London, Ontario. (unpublished results)

71 Leite[1], A.; Kemper[1], E. L.; Bonaccorsi[2], E. D.;. da Silva[1], M. J; Siloto[2], R.; El-Dorry[2], H.F.; Arruda[1], P. [1]Centro de Biologia Molecular e Engenharia GenTtica ; Universidade Estadula de Campinas, , Campinas, SP, [2]Instituto de Qufmica, Universidade de Spo Paulo, SP, Brazil. (unpublished results).

72 Anzai[1], H.; Takaiwa[2], F.;Katsumata[1], K. [1]Allergen Free-Technology (AFT) Laboratories Inc., [2]National Institute of Agrobiological Resources , c/o Meiji Seika Kaisha, Ltd., Morooka-cho, Kohoku-ku, Yokohama, Japan (unpublished results).

73 Arakawa T, Chong DK, Slattery CW, Langridge WH. Adv Exp Med Biol. 1999;464:149-59.

74 Huang[1], N.; Huang[1], J.;Wu[1], L.; Nandi[1],S;. Bartley[1],G.; Lonnerdal[2], Brodriguez[1,3],R. L. [1]Applied Phytologics Inc., Sacramento CA, [2]Department of Nutrition, UC Davis, [3]Section of Molecular and Cellular Biology, U.C. Davis, Davis CA (unpublished results).

75 Takase K, Hagiwara K. J Biochem (Tokyo). 1998 Mar;123(3):440-4.

76 Chong DK, Roberts W, Arakawa T, Illes K, Bagi G, Slattery CW, Langridge WH. Transgenic Res. 1997 Jul;6(4):289-96.

77 Terashima M, Murai Y, Kawamura M, Nakanishi S, Stoltz T, Chen L, Drohan W, Rodriguez RL, Katoh S. Appl Microbiol Biotechnol. 1999 Oct;52(4):516-23.

78 Sijmons PC, Dekker BM, Schrammeijer B, Verwoerd TC, van den Elzen PJ, Hoekema A. Biotechnology (N Y). 1990 Mar;8(3):217-21.

79 Fitchen J, Beachy RN, Hein MB. Vaccine. 1995 Aug;13(12):1051-7.

80 Torres[1], E.; Vaquero[2], C.; Stöger[1], E.; Sack[2], M.; Nicholson[1] L.; Drossard[2], J.; Christou[1], P.; Fischer[2], R.; Perrin[1], Y. [1]John Innes Centre,Norwich, UK, [2]RWTH Aachen, Institute for Biology, Aachen, Germany. (unpublished results).

81 Zeitlin L, Olmsted SS, Moench TR, Co MS, Martinell BJ, Paradkar VM, Russell DR, Queen C, Cone RA, Whaley KJ. Nat Biotechnol. 1998 Dec;16(13):1361-4.

82 Briggs[1], K.; Horn[5], M. E.; Fitchen[1], J.; Wang[1], F.; Zeitlin[2], L.; Love[5] R. T.; Jilka[5], J.; Burton[4], D.; Whaley[2], K. J.; [1]EPIcyte Pharmaceutical, Inc., San Diego, CA. [2]ReProtect, LLC, and [3]The Johns Hopkins University, Baltimore, MD. [4]The Scripps Research Institute, La Jolla, CA. [5]ProdiGene, Inc., College Station, TX. (unpublished results).

83 Ma SW, Zhao DL, Yin ZQ, Mukherjee R, Singh B, Qin HY, Stiller CR, Jevnikar AM. Nat Med. 1997 Jul;3(7):793-6.

84 Ma S, Jevnikar AM. Adv Exp Med Biol. 1999;464:179-94.

85 Bischoff, D.; Bates, M.; Welter,L. AgriVax, Inc. Los Angeles ,CA (unpublished results).

86 Weiner HL, Mayer LF. Ann N Y Acad Sci. 1996 Feb 13;778:xiii-xviii.

87 Weiner HL, Friedman A, Miller A, Khoury SJ, al-Sabbagh A, Santos L, Sayegh M, Nussenblatt RB, Trentham DE, Hafler DA. Annu Rev Immunol. 1994;12:809-37.

88 Hoyne GF, Lamb JR. Immunol Cell Biol. 1996 Apr;74(2):180-6.

89 Weiner HL, Mackin GA, Matsui M, Orav EJ, Khoury SJ, Dawson DM, Hafler DA. Science. 1993 Feb 26;259(5099):1321-4.

90 Sieper J, Kary S, Sorensen H, Alten R, Eggens U, Huge W, Hiepe F, Kuhne A, Listing J, Ulbrich N, Braun J, Zink A, Mitchison NA. Arthritis Rheum. 1996 Jan;39(1):41-51.

91 Bergerot I, Ploix C, Petersen J, Moulin V, Rask C, Fabien N, Lindblad M, Mayer A, Czerkinsky C, Holmgren J, Thivolet C. Proc Natl Acad Sci U S A. 1997 Apr 29;94(9):4610-4.

92 Plant-Derived Biologics Seminar and Public Hearing on Plant-Derived Biologics, April 5 & 6, 2000, Ames, Iowa USA.

93 Savoie,K. Edible vaccine success. Nature Biotechnology 2000, 18:367.

Chapter 15

Food Allergy: Recent Advances in Food Allergy Research

S. J. Maleki and B. K. Hurlburt

Department of Food Processing and Sensory Quality, Southern Regional
Research Center, Agricultural Research Service, U.S. Department
of Agriculture, 1100 Robert E. Lee Boulevard, New Orleans, LA 70124

Approximately 8% of children and 1-2% of adults have some
type of food allergy. Peanuts, fish, tree nuts, and shellfish
account for the majority of food hypersensitivity reactions in
adults, while peanuts, milk, and eggs cause over 80% of food
hypersensitivity reactions in children. Unlike the food hypersensitivity reactions to milk and eggs, peanut allergy is often severe,
persists into adulthood, lasts for a lifetime and seems to be
increasing in prevalence. Food allergies will be discussed as a
growing concern and a public health issue. In addition, the
current statistics, prevalence, and known symptoms of food
allergies will be addressed. Food allergy research lags far behind
the aeroallergen studies and the most advanced research in the
area of food allergies has been performed on peanuts. Therefore,
the final section of the chapter is dedicated to some of the most
recent research in the area of peanut allergy.

Adverse reactions to foods can be divided into three general groups: (i) allergy,
ii) intolerance, iii) toxicity. A true food allergy is characterized by a cascade of
immunological events involving antibodies, such as immunoglobulin E (IgE), and

IgG, cells of the immune system such as B-cells, mast cells and basophils, as well as, histamines and a host of other chemical mediators of inflammation. Intolerance to food is usually due to a deficiency of a metabolic enzyme and does not involve the immune system. For example, in the case of intolerance to milk an individual is deficient in lactase and is therefore unable to metabolize lactose. This disorder is usually an inherited trait, affecting up to 10% of the population. Food toxicity occurs due to ingestion of a toxic contaminant and does not have an immune basis.

Allergies to peanut, milk and egg account for over 80% of food hypersensitivity reactions found in children. The overwhelming majority of food allergies are classified as type I or IgE mediated, in which the antibody plays a major role in the cascade of immunological events that occur following ingestion of an allergenic food. Hypersensitivity or allergic reactions to milk and egg are often outgrown at an early age, whereas reactions to peanut, tree nuts, fish and shellfish can be life threatening, are rarely outgrown and appear to be increasing in prevalence. Coeliac disease, which is not IgE mediated, but nevertheless an allergic reaction, is also life long and requires the strict avoidance of gluten containing grains. It was shown that 90% of positive food allergic reactions in children where caused by eight foods now referred to as "the big eight" (1) The list of food allergens includes peanut, crustaceans, egg, fish, milk, soya bean, tree nuts, and wheat. There is no definitive treatment for food allergies other than avoidance, especially in cases where the reaction is quite severe.

Food allergies have become a major public health issue in many countries and have recently pre-occupied government agencies with the publication of official reports. For example, an expert committee has been convened by the UK Department of Health to make recommendations on peanut avoidance in women during pregnancy, lactation, and in children under the age of 3 (2). A recent publication in the Journal of American Medical Association demonstrated that intact peanut allergens were secreted in the breast-milk of 50% of lactating women. Based on this finding lactating women with family histories of allergies or genetic predisposition (referred to as "at risk" families) were advised to avoid peanuts during lactation (3). Currently, the issue concerning exposure of infants to food allergens through breast milk remains controversial and it is not clear whether it causes sensitization or tolerization of the infant. Another food allergy expert group has been formed by the International Life Sciences Institute (ILSI) to establish scientifically based criteria for deciding which food constituents should be labeled even if only a minute amount of a particular allergen is present (4). The allergenic foods listed by ILSI include the "big eight" with the inclusion of sesame seed. In addition to the increase in the prevalence of food allergies in the western countries, one of the major reasons for these concerns is the production and use of genetically modified organisms that may contain hidden allergens.

In the past several years a number of allergens have been identified that stimulate IgE production and cause IgE-mediated disease in man. Some of the immunological properties of allergenic proteins include their ability to bind serum specific IgE, elicit a positive prick skin test, and stimulate T-cell proliferation and the release of histamines from the mast cells of sensitive individuals. While the immunological characteristics of a number of food allergens have been determined, and despite increasing knowledge of the primary amino acid sequence of the identified allergens, specific features of the allergen contributing to IgE antibody formation have not been fully elucidated. Some known characteristics of allergens are that they are low molecular weight proteins or glycoproteins, which are abundant in the food source and stable to digestion by the gastrointestinal enzymes. These properties seemingly allow rapid penetration of undigested allergenic fragments at the mucosal membrane, facilitating sensitization and the immediate symptoms observed in allergenic patients.

Symptoms, causes and diagnosis of food allergy

Food allergies particularly cause problems for infants and young children with reported incidences of 6-8% in comparison to 1-2% in adults. A wide spectrum of clinical reactions is seen in allergic disease, the most prominent symptoms include cutaneous (89%), respiratory (52%) and gastrointestinal (GI, 32%) symptoms (7). Approximately 31% of these individuals have two symptoms and 21% have all three of the symptoms specified (7). Other less prevalent symptoms such as headache, sleeplessness, neurological problems (such as irritability, nervousness and mood swings) and abrupt changes in body temperature have also been attributed to food allergy. A single symptom or laboratory test is usually not enough to diagnose food allergy and in fact food allergies are often neglected or mis-diagnosed because the reactions can be complex and variable. For example, GI food allergy poses a challenge to the clinician because of its variable symptoms and lack of reliable diagnostic tests.

There are many accepted and some controversial factors that are considered to contribute to the development of allergy such as hereditary factors, the frequency and duration of breast feeding, GI problems, overall diet and nutrition, and environmental factors such as life style, cultural habits, in addition to levels, frequency and time or age of exposure to a particular allergen. For example, certain scientific reports suggest that the reduction in communicable disease as a result of increased hygiene and vaccinations in western nations is a major contributor to the development of higher incidence of allergic disease in these countries (8, 9).

Most often food allergy diagnosis is based upon a favorable response to an elimination diet and a positive response to a challenge with the suspected food. In the case of infants and children with more severe symptoms, the condition is treated by eliminating the suspected food from the diet for as long as 9-12 months such as in the case of cow's milk allergy. Approximately 2-3% of young children develop allergy or intolerance to cow's milk. In one study where a dietary survey was conducted to assess the nutrient intake of children on cow's milk restricted diet, cow's milk protein-free and reduced diets were compared to a group of cow's milk consumers (10). Significant differences were found in the nutrient intake of these groups. Children on milk-free diets had significantly lower intake of energy, fat, protein, calcium, riboflavin and niacin. An improvement was seen if milk substitutes were used, however the recommended levels of riboflavin and calcium were still not met. This study clearly shows that there is a risk of malnutrition in children deprived of cows' milk and possibly other foods, and that the parents of children on these types of avoidance diets need advice about food choices to reduce the chances of this occurring. When a child or adult is afflicted with more than one type of food allergy, or when a particular offending food is widely used in various food preparations and grocery store products, balancing between food avoidance and a healthy diet can be quite complex. In both adults and children, cases of multiple allergies cannot only lead to weight loss and malnutrition, but also have crippling physical and mental effects. The role of restricting the diet of mothers with food allergy during pregnancy or lactation is still under debate.

Cellular and humoral response in allergic disease

The development of an IgE response to an allergen involves a series of interactions between antigen-presenting cells (APCs), T cells, and B cells. When food is ingested, digested and absorbed into the blood stream, APCs acquire and then present peptide fragments (T-cell and B-cell epitopes) in conjunction with major histocompatibility complex (MHC) class II molecules to T-cells and B-cells. It is important to note that although the T-cell and B-cell epitopes may overlap, they are not necessarily the same. T-cells bearing the appropriate complementary T-cell receptor (TCR) will bind to the peptide-MHC complex on APCs leading to further interactions that result in the generation of a "second" (intracellular) signal, T-cell proliferation and cytokine secretion by T-cells. Cytokines are known to transmit signals between the various cells of the immune system. An inflammatory or non-inflammatory response depends on both the characteristics of the stimulus presented to the T cells via the APCs, as well as the type of cytokines that are secreted. IL4 is an inflammatory cytokine that is known to cause antibody class switching to promote IgE production and secretion. Part of the secreted IgE antibodies are known to become attached to the surface of mast cells and basophils *via* high affinity IgE binding receptors (FcεRI). When an allergen or a fragment of an allergen binds to and cross-links more than one IgE molecule it leads to clustering of the high affinity receptors. The clustering activates an intracellular signal transduction pathway, resulting in the degranulation of mast cells and basophils, and subsequent release of histamines and other chemical mediators of inflammation into the blood stream. This is why antihistamines are the classical form of drug relief for allergic symptoms. It is important to note that a fragment of an allergenic molecule must be large enough to contain more than one IgE binding site in order to cross-link two adjacent IgE molecules, cause histamine release and the symptoms of an allergic reaction. Opposing effects of various cytokines, such as IL-4 and IL-13 versus IFN-γ and TGF-β are involved in the regulation of IgE production by B-cells. Allergen specific T-cells, that are known to secrete high levels of cytokines, such as IL-4 and low levels of IFN-γ, are referred to as Th2 type T-cells, and play an important role in the production of IgE and the pathogenesis of allergic disease. The critical role T-cells play in this process has been studied in a variety of air-borne allergens (5), however, the role of T lymphocytes and antigen specificity in the induction and regulation of the food allergic response is less well defined. A recent review by Broide discusses the complexities of cellular and humoral responses involved in allergic disease more thoroughly (6).

Allergy to peanut

Hypersensitivity to peanut is a significant health problem and a great concern for the peanut industry. It has been estimated that 3 million Americans (1.1% of the population) are affected by peanut and/or tree nut allergies (11). Unlike hypersensitivity to most other foods such as egg or milk, hypersensitivity to peanut often persists throughout adulthood. Symptoms exhibited by patients with peanut hypersensitivity are often severe and can result in anaphylaxis and occasionally death (12, 13, 14). Currently, the most effective way to control peanut allergy is through complete elimination from the diet. However, peanuts are unusually common in the food supply, being found in an extraordinarily wide range of products. The most common cause of anaphylaxis due to foods results from accidental exposure to allergens. For example, leguminous products (i.e. peanuts) are often used as thickening agents in a variety of foods such as chili and stews. Chocolate containing minute amounts of peanut when it is labeled as non-peanut containing due to contamination of equipment in the production line, and inhalation of peanut dust on an airplane due to fellow passengers opening their peanut packets are some examples of accidental exposure to allergens. At least 55% of peanut allergic individuals have 1-2 accidental ingestions every 5.5 years (7).

Studies into the mechanism of peanut hypersensitivity have shown the existence of two major peanut allergens - Ara h 1, Ara h 2 and three minor allergens Ara h 3, Ara h 5 and Ara h 6 (15, 16, 17, 18). Ara h 4 was found to be a break down product of one of the previously identified allergens. An allergen is classified as a major allergen when it is recognized by greater than 90% of the individuals that are allergic to that particular food. Ara h 1, 2 and 3 are seed storage proteins belonging to the vicilin, conglutin, and glycinin families, respectively Ara h 5 has been identified as profilin and Ara h 6 remains to be identified as a unique allergen.

Each of the cDNAs for the allergens has been cloned and the nucleic acid and amino acid sequences determined (17, 18, 19, 20). The major IgE binding sites (or IgE epitopes) have been identified for Ara h 1, Ara h 2 and Ara h 3 and appear to be evenly distributed throughout the linear sequence of the molecules (17, 19, 21). Ara h 1 is 63 kDa and contains 23 IgE epitopes, while Ara h 2 is approximately 20 kDa with 10 IgE epitopes. Ara h 3 is 60 kDa, consists of an acidic and a basic subunit with 4 IgE binding sites located on the acidic, 40 kDa subunit. No common amino acid sequences or motifs have been found in any of the epitopes identified. In addition, it has been demonstrated that single amino-acid changes within each of the IgE binding epitopes of Ara h 1, Ara h 2 and Ara h 3 result in significant loss of IgE binding (reviewed in 22).

All of the IgE binding sites for Ara h 1, Ara h 2 and Ara h 3 have been mutated and the mutant recombinant proteins have been expressed. A single amino acid change in each epitope resulted in either elimination or a significant reduction of IgE binding to the mutated allergens (22). Hypothetically, these clones can be used for several purposes, such as T-cell immunotherapy and the engineering of a hypo-allergenic peanut plant. Whereas, there are complications and limitations for both of these approaches, active research towards these goals is underway.

The idea of generating a hypoallergenic plant necessitated finding functional assays for the allergens. While the allergens pose a substantial risk to sensitive individuals, they also play an important role in the development of peanut plants. Therefore, the goal is to change amino acid sequences in the IgE epitope regions to reduce the immune response, but retain enough native properties for normal protein folding, assembly and deposition in peanut seeds. Our first objective was to study the structure-function relationship of the individual allergens. A computer homology-based model of Ara h 1 was constructed (21). Fluorescence anisotropy and chemical cross-linking were utilized to show that Ara h 1 forms homotrimers that associate *via* strong hydrophobic interactions. In the quaternary structure the IgE epitopes of Ara h 1 are clustered in two main regions. The IgE epitopes of Ara h 1 were found to be clustered at the ends of each molecule, which coincides with the location of monomer-monomer contact and the strong hydrophobic intermolecular interactions. Given this, it seems unlikely that an Ara h 1 protein with mutations to reduce IgE binding to all 23 epitopes would yield a function protein in a putative hypoallergenic peanut plant. However, the structural studies do suggest a mechanism that may be responsible for the allergenicity of these peanut proteins. Due to the strong hydrophobic interactions between the monomers, it is possible that the IgE binding sites might be protected from digestive enzymes upon ingestion. This would in turn allow large fragments containing more than one IgE epitope to escape digestion and become absorbed into the blood stream. Large allergen fragments containing multiple epitopes are a requisite for IgE cross-linking and initiating the inflammatory response.

Recombinant, wild type Ara h 1 was unable to form trimers (A.W. Burks and G.A. Bannon, unpublished observation). Presumably, the *E. coli* produced, recombinant protein was not properly folded. Like other seed storage proteins, Ara h1 is presumably synthesized on rough endoplasmic reticulum and transits through this organelle to its destination in vacuolar protein bodies. As *E. coli* lacks these organelles, it is likely that additional assembly factors, or other post-translational modifications (such as glycosylation) are not present to help facilitate proper folding. The glycosylation found in the native Ara h 1 and missing in the recombinant protein may be responsible for improper folding. It is also possible that the glycosylation plays an important role in the quaternary structure formation. Since the recombinant Ara h 1 was impaired for trimer formation, the mutant recombinant Ara h 1 was not studied further. However, the resistance of Ara h 1 to digestive enzymes has been tested (23). One of the classical characteristics of food allergens is that they are more resistant to digestive enzymes than non-allergic food proteins. A mechanism or reason for this consistent observation had not been offered. As mentioned above, it is reasonable to hypothesize that the quaternary structure of Ara h 1 contributes to its resistance to digestion. To test this hypothesis, Ara h 1 was digested with three different enzymes: pepsin, the primary proteolytic enzyme in the stomach, and the intestinal proteases trypsin and chymotrypsin (23). Fragments of Ara h 1 as large as 58 kDa were found to survive digestion with the digestive enzymes over a three hour period. After 24 hours fragments between 20 - 30 kDa were still visible. When the digestion resistant fragments were isolated and sequenced, they were found to contain the immuno-dominant IgE binding epitopes of Ara h 1.

Peanut allergenicity and processing

Very few studies have addressed the effects of processing on the allergenic properties of foods. Thermal processing such as roasting, curing and various types of cooking, can cause several non-enzymatic, biochemical reactions to occur in foods (24). One of the major reactions that occurs during cooking or browning of foods is known as the Maillard reaction (25, 26), which is important in the development of flavor and color in peanuts as well as many other processes of the food industry. The amino groups of proteins are modified *via* reducing sugars to form Schiff's bases that can undergo rearrangement to form "Amadori products" (Figure 1). Subsequently, the Amadori products are degraded into dicarbonyl intermediates. These intermediary compounds that are more reactive than the parent sugars (with respect to their ability to react with amino groups of proteins) and form cross-links, or stable end products called advanced Maillard reaction products (MRPs) or advanced glycation end products (AGEs). It is known that in addition to cross-linking, advanced Maillard reactions could lead to the loss or modification of amino acids such as lysine (see carboxymethyllysine or CML in Figure 1), malanoidin formation, and other non-cross-linking modifications to proteins that may have detrimental nutritional, physiological, and toxicological consequences (25).

A collection of studies have addressed the immunological recognition and responses to the advanced MRPs (or AGEs). The protein products modified by the Maillard reaction have been shown to evoke an immunoglobulin G (IgG) response, which has been correlated by a number of studies to IgE production (28-31). AGEs have also been shown to promote monocyte migration (32, 33) and the production of cytokines (34). AGEs are associated with heightened immunogenicity, aging, and age-enhanced disease states such as diabetic complications, atherosclerosis, hemodialysis-related amyloidosis, and Alzheimer's disease. However, few studies have been done to address the role of these products on the allergenic properties of ingested foods (27, 35-39).

Our studies revealed that roasted peanut extracts bound serum IgE from allergic individuals at significantly higher levels than the raw peanut extracts (27). In order

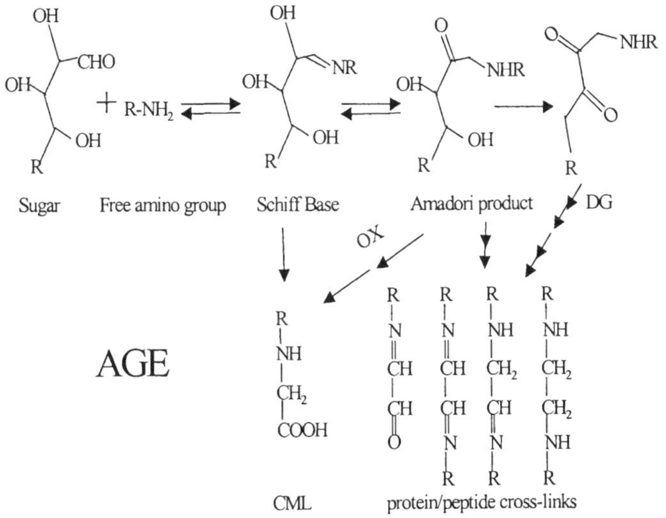

Figure 1. The Maillard reaction and the formation of the advanced glycation end products (AGE). An example of a glucasone: 4 deoxy glucasone (DG) is shown above. Reproduced from reference 27.

to understand enhanced immunogenicity by roasted peanuts and the contribution of the Maillard reaction to IgE binding properties, susceptibility of the major peanut allergens to digestion by gastrointestinal fluid and their heat stability were assessed. A large number of biochemical modifications to proteins are known to occur during roasting/browning of foods. Therefore the allergens were purified from raw peanuts and used in a well documented (25, 26), highly characterized, and isolated, *in vitro* model system to determine if the Maillard reaction alone affects the allergenic properties of these allergens. In this model system, referred to as the simulated roasting model (SRM), proteins were incubated in the presence of sugars, heated over time and assessed for allergenic properties. After whole peanut proteins, Ara h 1 and Ara h 2 from raw peanuts were subjected to the SRM, they became more resistant to digestion with gastric fluid, less soluble, underwent structural modifications, and bound higher levels of IgE. Deciphering the biophysical modifications to the allergens following the SRM and the implication of these changes on the immunological properties of peanut proteins was the next logical step. In the SRM, intermolecular cross-links were formed between Ara h 1 monomers to generate covalently associated trimers and hexamers and although Ara h 2 did not form higher order structures, it was modified by intramolecular cross-links that rendered it highly resistant to digestive enzymes. As previously discussed, Ara h 1 from raw peanuts has been shown to form stable trimeric complexes in solution at low concentrations, which is suggested to play a role in the allergenic properties of this protein. The formation of a trimeric complex may allow the molecule some protection from protease digestion and denaturation, allowing passage of large fragments of Ara h 1 containing several intact IgE binding sites across the lumen of the small intestine, therefore, contributing to its allergenicity. The reversible association of Ara h 1 monomers through hydrophobic interactions, previously hypothesized to be important in allergenicity, becomes an irreversible covalent cross-linking due to thermal processing. Therefore, Ara h 1 subjected to the SRM becomes more resistant to digestive enzymes than previously determined for unmodified Ara h 1 purified from raw peanut extracts. In turn, this affects its overall ability to induce an allergic response.

Nordlee *et al.* reported that roasted peanuts bind IgE at higher levels than raw peanuts and that the IgE recognition sites in roasted peanuts differ from those of raw peanuts (38). They hypothesized that these observations may be due to the fact that heat treatment increases the allergenicity of peanut proteins by increasing the availability of allergic binding sites on the proteins that were previously unexposed. Our findings imply that in addition to exposing previously unavailable sites, the covalent modification of the proteins during the roasting process may create novel IgE binding sites and enhance other allergenic properties such as resistance to heat, degradation and digestion by gastric secretions.

To test the findings of our SRM, whole peanuts were roasted for various lengths of time and the allergens were compared for solubility, IgE binding and induction of T cell proliferation (39). In addition, the biophysical and immunological properties of Ara h 1 and Ara h 2 purified from a medium roast peanut samples and from raw peanuts were compared (39). The whole peanut extracts from the various roasted peanut samples showed decreasing solubility and T cell proliferation, with expansive structural changes to the proteins and heightened IgE binding with increased time of roasting, indicating a dominant role of the Maillard reaction. In addition, cross-linked Ara h 1 trimers and hexamers, as well as Ara h 2 protein (similar to the protein following the SRM) were found in purified fractions. Antibodies against some AGE bi-products, such as anti-CML, hydroxynonenol (HNE), and malondialdehyde (MDA) were used to determine if any of these specific modifications contribute to the increase in IgE binding (26). The level of CML modifications to the allergens was found to correlate with the increase in IgE binding. Based on these findings we conclude that thermal processing events can drastically alter the biophysical and immunological properties of proteins. If processing can cause an increase in allergenic properties of proteins, steps should be taken to minimize these reactions during the roasting process. Our laboratory is in the process of investigating the effects of Maillard reaction inhibitors on the allergenic properties of peanut proteins and novel processing methods that can reduce these properties.

References

1. Bock, A.A., Sampson, H., Atkins, F.M., et al. Double blind placebo controlled food challenge (DBPCFC) as an office procedure: a manual. J. Allergy Clin, Immunol. 1988, 82, 986-97.
2. Committee on Toxicity of Chemicals in Food. Consumer Products and the Environment. Peanut Allergy: London Department of Health 1998, 1-57.
3. Vadas P., Wai, Y., Burks, A.W. Perelman, B. Detection of peanut allergens in breast milk for lactating women. JAMA. 2001, 285(13),1746-8.
4. Bousquet, J., Bjorksten, B., Bruinzeel-Kosmen, C.A., et al. Scientific criteria and the selection of allergenic food for product labeling. Allergy 1998, 53, 3-21(supplement).
5. O'Hehir, R.E., Garman, R.D., Greenstein, J.L., and Lamb, J.R. The specificity and regulation of T-cell responsiveness to allergens. Annu. Rev. Immonol. 1991, 9, 67-95.
6. Broide, D. Molecular and cellular mechanisms of allergic disease. J. Allergy Clin. Immunol. suppl. 2001, 108 (2), S65-71.
7. Sicherer, S., Burks, A.W., and Sampson, H.A. Clinical features and acute allergic reactions to peanut and tree nut in children. Pediatr. 1998, 102 (1), 131-132.
8. Voelker, R. The hygiene hypothesis. JAMA. 2000, 283, 1282.

9. Lynch, N.R., Goldblatt J., Le Souef P.N. Parasitic infections and risk to asthma and atopy. Thorax 1999, 54, 659-60.

10. Henriksen, C., Eggesbo, M., Halvorsen, R., Botten, G., Nutrient intake among two-year-old children on cow's milk restricted diets. Acta paediatr. 2000, 89(3), 272-8.

11. Sicherer, S., Monoz-Fulong, A., Burks, A.W., and Sampson, H.A. Prevalence of peanut and treenut allergy in the US population by a random digit dial telephone survey. J. Allergy Clin. Immunol. 1999, 103 (4), 449-562.

12. Fries, J.H. Peanuts: allergic and other untoward reactions. Annal. Allergy 1982, 48(4), 220-6.

13. Sampson, H., Mendelson, L., Rosen, J.P. Fatal and near fatal anaphylatic reactions to food in children and adolescents. N. Engl. J. Med. 1992, 327(6), 380-348.

14. Yunginger, J.W., Sweeney K.G., Surner, W.Q., Giannandrea, L.A., Teigland, J.D., Bray, P.Q., York, J.A., Beidrzycki, L., Squillance, D.L., et al. Fatal food-incuded anaphylaxis. JAMA, 1988, 260 (10), 1450-2.

15. Burks, A.W., Williams, L.W., Connaughton, C., Cockrell, G., O'Brien, T.J., Helm, R.M. Identification of a major peanut allergen, Ara h I, in patients with atopic dermatis and positive peanut challenges. A. Allergy Clin. Immunol. 1992, 90 (6 Pt. 1), 962-9.

16. Burks, A.W. Williams,, L.W., Helm, R.M., Connaughton, C., Cockrell, G., O'Brien, T.J., Identification and characterization of a second peanut allergen, Ara h II, with use of the sera of patients with atopic dermatitis and positive peanut challenge. J. Allergy Clin. Immunol. 1991, 88(2), 172-9.

17. Rabjohn P., Helm R.M., Stanley J.S., West C.M., Sampson H., Burks A.W., et al. Molecular cloning and epitope analysis of the peanut allergen. Ara h 3. J. Clin. Invest. 1999, 103, 535-52.

18. Janke, T.K., Crameri, R., Appenzeller, U., Schlakk, M., Becker, W-M. Selective cloning of peanut allergens, including profilin and 2S albumins, by phage display technology. Allergy Immunol. 1999, 119, 265-274.

19. Stanley, J.S., King, N., Burks, A.W., Huang, S.K., Sampson, H., Cockrell, G., et al. Identification and mutational analysis of the immunodominant IgE binding epitopes of the major peanut allergen Ara h 2. Arch. Biochem. Biophys. 1997, 342-244-53.

20. Burks, A.W., Cockrell G., Stanley J., Helm R., Bannon G.A. Recombinant peanut allergen, Ara h 1, expression and IgE binding in patients with peanut hypersensitivity. J. Clin. Invest. 1995, 96, 1715-21.

21. Shin D., Compadre C.M., Maleki S.J., Kopper R.A., Sampson H., Huang S.K., et al. Biochemical and structural analysis of the IgE binding sites on Ara h 1, an abundant and highly allergenic peanut protein. J. Biol. Chem. 1998, 273, 13753-59.

22. Burks A.W., Sampson H.A., Bannon G.A. Peanut allergens. Allergy 1998, 53, 725-30.

23. Maleki S.J., Kopper R.A., Shin, D.S., Stanley S.J., Sampson H., Burks A.W., Bannon G.A. Structure of the major peanut allergen Ara h 1 may protect IgE-binding epitopes form degradation. J. Immunol. 2000, 164, 5844-49.

24. Shahidi, F., Ho, T-C. Process-Induced Chemical Changes in Foods, New York (NY) Plenum Press;1998.

25. Maillard, L.C. Action des acides amines sur les sucres: formation des melanoidines par voie methodique, C.R. Acad.Sci. 1912, 154, 66-8.

26. Maillard, L. C. Formation d'humus et de combustibles mineraux sans intervention de l''oxygiene atmospherique, de micoorganismes, de hautes temperatures, ou des fortes pressions. C. R. Acad. Sci. 1912, 155, 1554-8.

27. Maleki, S.J., Chung S-Y., Champagne, E.T., Raufman, J-P. The effects of roasting on the allergenic properties of peanut proteins. J. Allergy. Clin. Immunol. 2000, 106, 763.

28. Campbell, D.E., Ngamphaiboon, J., Clark, M.M. Indirect enzyme-linked immunosorbent assay for measurement of human IgE and G to purified cow's milk proteins: application in diagnosis of cow's milk allergy J. Clin. Microbiol. 1987, 25, 2114-19.

29. Heddleson, R. A., Park, O., Allen, J.C. Immunogenicity of casein phosphopeptides derived from tryptic hydrolysis of β-casein. J. Dairy. Sci. 1997, 80, 1971-76.

30. Duchateau, J., Michilis, A., Lambert, J., Cossart, B., Casimir, G. Anti-β-lactoglobulin IgG antibodies bind to a specific profile of epitopes when patients are allergic to cow's milk proteins. Clin. Exp. Allergy 1998, 28, 824-33.

31. Ikeda, K., Higashi, T., Sano, H., Jinnouchi, Y., Yoshida, A., Ueda, S., Horiuchi S., N-ϵ-(carboxymethyl) lysine protein adduct is a major immunological epitope in proteins modified with advanced glycation end products or the Maillard reaction. Biochemistry 1996, 35, 8075-83.

32. Kirstein, M., Brett, J., Radoff, S., Ogawa, S., Stern, D., Vlassara, H. Advanced protein glycosylation induces selective transendothelial human monocyte chemotaxis and secretion of PDGF: Role in vascular disease of diabetes. Proc. Natl. Acad. Sci. USA 1990, 87, 9010-14.

33. Schmidt, A., Yan, S., Brett, J., Mora, R., Nowygrad, R., Stern, D. Regulation of human mononuclear phagocyte migration by cell surface-binding proteins for AGE products. J Clin. Invest. 1993, 91, 2155-68.

34. Vlassara, H., Brownlee, M., Manogue, K.R., Dinarello, C.A., Pasagian, A. Cachectin/TNF and IL-1 induced by glucose-modified proteins: Role in normal tissue remodeling. Science 1988, 240, 1546-8.

35. Deshpande, S.S., Nielsen, J. In vitro digestibility of dry bean (Phaseolus Vulgaris L.) Proteins: The role of heat stable protease inhibitors. Food Sci. 1987 a, 52, 1330-34.
36. Deshpande, S.S., Nielsen, J. In vitro enzymatic hydrolysis of Phaseolin, the major storage protein of Phaseolus Vulgaris L. Food Sci, 1987 b 52, 1326-29.
37. Burks, A.W., Williams, L.W., Thresher, W., Connaughton, C., Cockrell, G., Helm, R.M. Allergenicity of peanut and soybean extracts altered by chemical or thermal denaturation in patients with atopic dermatitis and positive food challenges. J. Allergy Clin. Immunol., 1992, 90, 889-97.
38. Nordlee, JA, Taylor SL, Jones RT, Yunginger JW. Allergenicity of various peanut products as determined by RAST inhibition. J. Allergy and Clin. Immunol. 1981, 68, 376-82.
39. Maleki, S.J., Chung, S-Y., Champagne, E.T., Khalifah, R.C. Allergic and biophysical properties of peanut proteins before and after roassting. Food Allergy and Intoler., A Journal for the World Food Industry 2001, 2 (3), 211-221.

Chapter 16

Assessment of the Allergenicity of Foods Produced through Agricultural Biotechnology

S. L. Taylor

Food Allergy Research and Resource Program, 143 Food Industry Complex, University of Nebraska, Lincoln, NE 68583

Agricultural biotechnology provides an accurate and precise method for enhancing the beneficial traits of foods produced from plants, animals, and microorganisms. While traditional plant breeding, for example, results in the transfer of hundreds of genes and their proteins, agricultural biotechnology approaches allow the selection of single genes or groups of genes and their products. Whether by traditional plant breeding or by modern agricultural biotechnology, novel proteins can be introduced into the edible portion of the new plant variety. Historically, with traditional plant breeding the safety of these novel proteins has not been carefully assessed and would be extremely difficult since the number and precise identity of the novel proteins is typically unknown. However, with foods produced through modern agricultural biotechnology, only a limited number of well defined novel proteins are introduced into the new plant variety. This offers the opportunity for the assessment of the safety of these novel proteins. It must be emphasized that this is just one part of the overall safety assessment process for a food produced through agricultural biotechnology.

Initially, the safety assessment of foods produced through agricultural biotechnology focuses on an evaluation of their compositional equivalence to the parental variety/strain or to other commercial varieties. Compositional comparisons focus on key nutrients and anti-nutrients, naturally-occurring toxicants including allergens, and any trait-specific substances. With respect to allergenicity, the introduction of novel genes through modern agricultural biotechnology would not be expected to have much effect upon the inherent allergenicity of the recipient agricultural crop in most cases. For example, glyphosate-resistant soybeans have been evaluated for levels of various naturally-occurring toxicants found in soybeans including trypsin inhibitors and allergens. No differences were found in the levels of these toxicants in the genetically engineered soybeans by comparison to the parental soybean variety (*1*). Of course, agricultural biotechnology does offer the promise of reducing or eliminating allergenic proteins from existing foods. However, no such crops are currently being marketed.

A more important focus of safety of assessment of foods produced through agricultural biotechnology is the assessment of the allergenicity of the novel proteins introduced into the new plant variety. The allergenicity of the newly introduced protein has become a source of some concern in the safety evaluation process. Virtually all allergens are proteins (*2*). However, only a very few of the many proteins in nature are allergens so the inherent risk of allergenicity from any randomly selected protein is rather small. Allergens are those proteins which can induce the production of allergen-specific immunoglobulin E (IgE) in susceptible individuals (*3*). Even in commonly allergenic foods, only a few of the many proteins have the capability of inducing allergic sensitization in susceptible individuals. In nature, allergens can be found in many foods, pollens, mold spores, insect venoms, and other sources. Therefore, an assessment of the allergenicity of the novel protein should be made regardless of the source of the novel genes. Obviously, the risk of transfer of an allergenic protein is much greater when the DNA source is derived from a known allergenic source, although it is quite possible to transfer a non-allergenic protein from a known allergenic source. However, the possibility must also be considered that a potentially allergenic protein that does not cause much allergy because its expression is quite low, could present a larger problem if it is transferred to another species and expressed at a much higher level.

In IgE-mediated food allergies, allergen-specific IgE antibodies are produced in susceptible individuals by B lymphocytes in response to the immunological stimulus created by exposure of the immune system to the allergen (*4*). Food allergens are usually naturally-occurring proteins present in the food (*2*). Naturally-occurring proteinaceous allergens, whether from foods or pollens, mold spores, animal dander, dust mites, insects, insect venoms, and other sources are those proteins that are capable of eliciting allergic sensitization. In the sensitization phase, a susceptible individual forms allergen-specific IgE antibodies after exposure to a specific food or environmental protein. The IgE antibodies, including the allergen-specific IgE antibodies, bind to the surfaces of mast cells in the tissues and basophils in the blood. The sensitization phase is asymptomatic. Upon subsequent exposure of to the specific allergen, the allergen cross-links two or more of the allergen-specific IgE antibodies affixed to the surfaces of mast cells or basophils (Figure 1). This interaction triggers the disruption of the mast cell/basophil membrane and the release of a variety of potent physiologically active mediators into the bloodstream and tissues. The granules within mast cells and basophils contain many mediators of the allergic

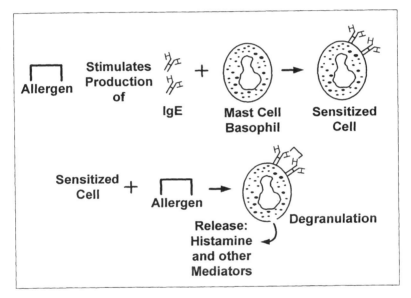

Figure 1. Mechanism of IgE-Mediated Allergic Reaction

reaction. It is the interaction of these mediators, such as histamine, with a variety of receptors in different tissues that is responsible for provoking an allergic reaction. As an example of the effects of just one of several dozen mediators released from mast cells and basophils, histamine can elicit inflammation, pruritis, and contraction of the smooth muscles in blood vessels, gastrointestinal tract, and respiratory tract (5). IgE-mediated allergies are called immediate hypersensitivity reactions because exposure to the allergen provokes symptoms within a few minutes to one or two hours after exposure. In the case of IgE-mediated food allergies, numerous symptoms ranging from mild and annoying to life-threatening can develop (Table I). The nature and severity of the symptoms can vary between individuals, between one episode and another, and can depend upon the frequency and dose of exposure to the offending allergen.

Table I. Typical Symptoms of IgE-Mediated Food Allergies

Urticaria (hives)	Diarrhea
Asthma	
Dermatits	Vomiting
Rhinitis	
Angioedema	Nausea
Laryngeal edema	
Pruritis	Abdominal pain
Anaphylactic shock	

Even among susceptible individuals, it must be emphasized that exposure to specific food proteins, regardless of source, does not usually result in the formation of IgE antibodies. Typically, exposure to food proteins in the gastrointestinal tract results in oral tolerance through either the formation of protein-specific IgG, IgM, or IgA antibodies or no immunological response whatsoever (clonal anergy) (6). Thus, the probability of allergic sensitization to a specific novel protein in foods produced through biotechnology is rather low. Still, it is important to assess the potential allergenicity of all novel proteins introduced through agricultural biotechnology.

The overall prevalence of IgE-mediated food allergies is not precisely known. For all age groups, the prevalence of IgE-mediated food allergies is likely in the range of 2.0 – 2.5% (7). The prevalence rate for food allergies among infants is several times higher than among adults (5). However, the prevalence of environmental allergies to pollens, mold spores, animal danders, dust mites, etc.

is much higher affecting affecting perhaps as much as 15 – 20% of the overall population. Based upon these prevalence estimates, the assessment of the allergenicity of foods produced through agricultural biotechnology should be evaluated as a routine feature of the safety assessment process.

The assessment of the potential allergenicity of a food produced through agricultural biotechnology has two essential features. First, it must be determined if one of the novel proteins is an allergen derived from the donor organism, a protein to which some consumers would already be sensitized. If the novel gene is obtained from a known allergenic source, then the gene product must be assumed to be an allergen unless proven otherwise. Appropriately, the U.S. Food and Drug Administration (FDA) has focused their concern on the presence of DNA from known, especially commonly, allergenic sources in the genetically modified crop (8). Secondly, it must be determined if one of the novel proteins has the ability to elicit allergic sensitization. This is especially important in cases where the novel gene is derived from sources with no history of allergenicity. This part of the allergenicity assessment is particularly difficult because there is no single validated approach to prediction of the ability of a novel protein to sensitize susceptible consumers. As a result, a decision tree strategy involving several approaches has been devised to strengthen the predictive accuracy of the assessment.

Several years ago, the International Food Biotechnology Council (IFBC) in collaboration with the Allergy & Immunology Institute of the International Life Sciences Institute (ILSI) developed such a decision tree approach to the assessment of the potential allergenicity of genetically modified foods (9). Recently, this approach was modified slightly in the recommendations of a consultation conducted by the World Health Organization/Food & Agriculture Organization (Fig. 2, 10) This decision-tree strategy focuses on the source of the gene, the sequence homology of the newly introduced protein to known allergens, the immunological reactivity of the newly introduced protein with IgE from the blood serum of individuals with known allergies to the source of the transferred genetic material, and the physicochemical properties of the newly introduced protein.

The sources of genetic material can be classified as commonly allergenic, less commonly allergenic, or unknown allergenic potential. Commonly allergenic foods include milk, eggs, peanuts, soybeans, tree nuts, fish, crustacea, and wheat (11). These few foods probably account for more than 90% of all food allergies on a worldwide basis. The IFBC-ILSI report contains an extensive list of more than 160 foods and food-related substances that have been associated with allergic reactions in individuals (12). While some of these foods would appropriately be classified as less commonly allergenic foods, some judgment is required because these reactions were not always well investigated. Similarly if genes were obtained from sources of well known environmental allergens, such as ragweed pollen, these sources would have to be treated as commonly allergenic.

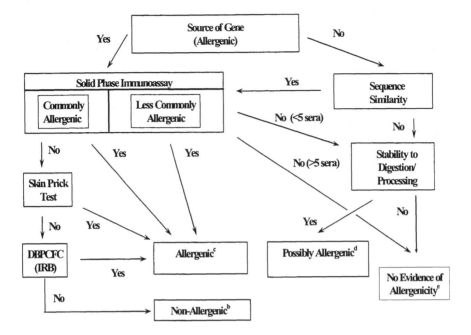

Figure 2. Decision-Tree Strategy for the Assessment of the Allergenicity of Genetically-Modified Foods (WHO, 2000)

(a) The figure was adapted from decision-tree approach developed by International Food Biotechnology Council and Allergy and Immunology of the International Life Sciences Institute (Metcalfe et al., 1996).

(b) The combination of tests involving allergic human subjects or blood serum from such subjects would provide a high level of confidence that no major allergens were transferred. The only remaining uncertainty would be the likelihood of minor allergen affecting a small percentage of the population allergenic to the source material.

(c) Any positive results obtained in tests involving allergenic human subjects or blood serum from such subjects would provide a high level of confidence that the novel protein was a potential allergen. Foods containing such novel proteins would need to be labeled to protect allergic consumers.

(d) A novel protein with either no sequence similarity to known allergens or derived from a less commonly allergenic source with no evidence of binding to

IgE from the blood serum of a few allergic individuals (<5), but that is stable to digestion and processing should be considered a possible allergen. Further evaluation would be necessary to address this uncertainty. The nature of the tests would be determined on a case-by-case basis.

(e) A novel protein with no sequence similarity to known allergens and that was not stable to digestion and processing would have no evidence of allergenicity. Similarly, a novel protein expressed by a gene obtained from a less commonly allergenic source and demonstrated to have no binding with IgE from the blood serum of a small number of allergic individuals (>5 but <14) provides no evidence of allergenicity. Stability testing may be included in these cases. However, the level of confidence based on only two decision criteria is modest. The Consultation suggested that other criteria should also be considered such as the level of expression of the novel protein.

With proteins derived from commonly allergenic and less commonly allergenic sources as well as sources of unknown allergenic potential, a comparison of the amino acid sequence of an introduced protein with the amino acid sequence of known allergens is a useful initial approach in the determination of allergenic potential (9). Potentially significant sequence homology would require a match of at least 8 contiguous identical amino acids (13; 9). This peptide length is thought to be the minimum size for a T cell-binding epitope. B cell epitopes are thought to be even larger. The IFBC-ILSI report contains a list of 198 sequences of food and environmental proteins that are reported to be allergens (9). Even more known allergen sequences are now known.

The next step in the evaluation of the allergenic potential of foods derived from genetically engineered plants involves an assessment of the immunoreactivity of the newly introduced proteins with IgE antibodies from the sera of individuals allergic to the donor plant. With proteins derived from known allergenic sources, the identification of individuals with well documented allergic reactions to foods derived from the donor plant is necessary. Blood serum containing allergen-specific IgE is obtained from individuals with such allergies to the donor plant, and the serum is tested for reactivity with the newly introduced protein or extracts of the transgenic food using immunoassays such as the radioallergosorbent test or RAST (14). A positive test certainly raises concerns about the allergenicity of the newly introduced protein. Unless these concerns can be convincingly discounted by additional in vivo testing using allergic subjects, such as skin prick testing or double-blind, placebo-controlled food challenges (9), foods containing the newly introduced gene should be considered as allergenic and appropriately labeled before being placed on the market. Obviously, the need for some, as yet undefined form of labeling and the need for careful segregation of such crops would be a major deterrent to their entry into the marketplace. Thus far, no such products have entered the marketplace.

When negative or equivocal results are obtained in the in vitro immunoassays, the genetically modified food should be investigated further using in vivo skin-prick tests with an appropriate number of allergic test subjects (15). The skin-prick test provides an in vivo indication of the allergenicity of the genetically modified food. The ultimate test of the potential allergenicity of the genetically modified food product would be the double-blind, placebo-controlled food challenges with allergic individuals (16). If evidence of allergenicity was obtained in either of these in vivo procedures, the foods containing the newly introduced gene should be considered as allergenic.

It is more difficult to identify individuals who are allergic to the less commonly allergenic foods. Since fewer sera would likely be available for immunoreactivity assessments, the approach for products containing novel proteins from such sources is a combination of immunoreactivity and physiochemical stability. Allergens are usually proteins that are stable to digestion (*13*; *17*). Thus, an easily digestible protein is much less likely to be capable of inducing allergic sensitization. The enzyme transferred into soybeans to make them tolerant of the herbicide, glyphosate, is rapidly digested in vitro (*18*) and is therefore unlikely to induce allergic sensitization. The use of digestive stability is certainly not a foolproof criterion. In all likelihood, some food proteins that are stable to digestion are probably not allergenic, although the digestive stability of food proteins has not been well cataloged. Certainly, some food proteins that are unstable to digestion are capable of eliciting a mild allergic reaction known as the oral allergy syndrome (*2*).

In assessing the possible allergenicity of novel proteins introduced through agricultural biotechnology, the most difficult assessment involves novel proteins obtained from sources of unknown allergenic potential such as viruses, bacteria, insects, non-food plants, and other non-food sources. Such substances are sometimes found in foods at very low levels, but they are not considered food sources. The likelihood that such proteins will be allergens is not very high because most proteins in nature do not stimulate allergic sensitization. Additionally, such novel proteins will often be expressed in the genetically modified food at very low levels, while allergic sensitization is more likely to occur to the major proteins that exist in foods. The modified approach to allergenicity assessment recently developed by WHO/FAO (*10*) suggests that the level of expression of the novel protein in the genetically modified food might be another factor to consider in the allergy assessment because those novel proteins expressed at very low levels would present little, if any, risk of allergic sensitization. However, the use of the level of expression as a criterion would be rather difficult because little information exists on the minimal threshold doses for allergenic proteins needed to elicit sensitization or to induce a reaction in a previously sensitized individual. The assessment of the allergenicity of proteins obtained from sources of unknown allergenic potential as outlined in the ILSI/IFBC approach (*9*) included evaluations of the amino acid sequence homology to known allergens and the physicochemical stability, especially the digestive stability of the protein (*13*; *9*).

In the ILSI/IFBC approach, if the gene of interest is obtained from a source with no history of allergy and the novel protein has no significant sequence homology with known allergens and is not stable to digestion, then it is concluded that the novel protein has little likelihood of eliciting allergic sensitization (*9*). In the more recent WHO/FAO approach, the level of expression of the novel protein in the genetically modified food and the functional classification of the novel protein are additionally considered (*10*; *11*). Many of the common plant-derived allergens fall into several selected functional categories, often various types of pathogenesis-related proteins (*19*). If a novel protein in a genetically modified food falls into one of the classes of commonly allergenic, plant-derived proteins, then it likely deserves careful scrutiny. It may likely have sequence homology to known allergens in that functional category, but even if it does not, it should be carefully assessed. The use of immunoreactivity using serum from individuals allergic to other proteins in the same functional category would be one approach to consider for such assessments.

The ILSI/IFBC approach has proven useful during the period of time that has elapsed since it was proposed. However, scientists and clinicians learn more about food allergy, the ability to predict the allergenicity of novel proteins should improve. Thus, this decision tree should properly be viewed as a dynamic approach that will be altered over time as more knowledge and better methods become available. In fact, a new decision tree strategy has been developed by FAO/WHO since the presentation that has lead to the development of this manuscript. This new decision tree strategy takes a more rigorous approach and uses targeted serum screening with serum from individuals allergic to related materials and animal testing in addition to the approaches advocated in the earlier ILSI/IFBC approach (20). The level of expression of the novel protein is not used as a criterion in this new approach because of the uncertainties regarding threshold doses for sensitization and reaction (20).

The approach advocated by the ILSI/IFBC decision tree and later related approaches have already proven useful. Some years ago, Pioneer Hi-Bred International succeeded in introducing a high-methionine protein from Brazil nuts into soybeans to correct the inherent methionine deficiency of soybeans. These soybeans would have enhanced value for animal feeding purposes. However, since Brazil nuts are known to be allergenic (21), the possible allergenicity of these genetically modified soybeans and the high-methionine protein was evaluated using RAST inhibition with the sera of Brazil nut-allergic individuals and skin-prick tests of such individuals. The high-methionine protein from Brazil nuts was demonstrated to be the major Brazil nut allergen (22). As a result, Pioneer Hi-Bred International decided not to commercialize this novel soybean variety. This example demonstrates that the decision-tree approach described above works in the assessment of the allergenicity of transgenic foods.

Agricultural biotechnology will ultimately affect many of the products currently in commercial markets. These novel products must be evaluated for their safety. However, many of the foods produced through agricultural biotechnology will be altered only slightly in composition from traditional foods. This is especially true for the first generation of such products that is reaching the market such as insect-resistant and herbicide-tolerant crops. Thus, the safety evaluation should be focused on the compositional differences especially the safety of any newly introduced proteins. The allergenicity of these newly introduced proteins should be one element of the safety assessment of these newly introduced proteins. A decision-tree strategy has been developed for the assessment of the allergenicity of genetically modified foods and has been shown to be useful in such evaluations.

References

1. Padgette, S.R.; Taylor, N.B.; Nida, D.L.; Bailey, M.R.; MacDonald, J.; Holden, L.R.; Fuchs, R.L. J. Nutr. 1996, 126, 702-716.
2. Bush, R. K.; Hefle, S. L. Crit. Rev. Food Sci. Nutr. 1996, 36, S119-S163.
3. Taylor, S. L.; Nordlee, J. A.; Bush, R. K. In *Food Safety Evaluation*; Finley, J. W.; Robinson, S. F.; Armstrong, D. J. Eds.; Am. Chem. Soc.: Washington D. C., 1992, pp 316-329.
4. Mekori, Y. Crit. Rev. Food Sci. Nutr. 1996, 36, S1-S18.
5. Lemke, P. J.; Taylor, S. L. In *Nutritional Toxicology*; Kotsonis, F. N.; Mackey, M.; Hjelle, J. Eds.; Raven Press: NY, 1994, pp 117-137.
6. Strobel, S. In *Food Allergy – Adverse Reactions to Foods and Food Additives*, 2nd ed.; Metcalfe, D. D.; Sampson, H. A.; Simon, R. A. Eds; Blackwell Science: MA, 1997, pp 107-135.
7. Taylor, S. L.; Hefle, S. L.; Munoz-Furlong, A. Nutr. Today 1999, 34, 15-22.
8. Food & Drug Administration. Fed. Reg. 1992, 57, 22984-23005.
9. Metcalfe, D. D.; Astwood, J. D.; Townsend, R.; Sampson, H. A.; Taylor, S. L.; Fuchs, R. L. Crit. Rev. Food Sci. Nutr. 1996, 36, S165-S186.
10. World Health Organization. Report of a Joint FAO/WHO Expert Consultation on Foods Derived Through Biotechnology, Geneva, Switzerland, May 29-June 2, 2000, 35 pp.
11. Food & Agriculture Organization. Report of the FAO Technical Consultation on Food Allergies, Rome, Italy, Nov. 13-14, 1995, 55 pp.
12. Hefle, S. L.; Nordlee, J. A.; Taylor, S. L. Crit. Rev. Food Sci. Nutr. 1996, 36, S69-S89.
13. Fuchs, R. L.; Astwood, J. D. Food Tech. 1996, 50, 83-88.
14. Yunginger, J. W.; Adolphson, C. R. In *Manual of Clinical Laboratory Immunology*, 4th ed.; Rose, N. R.; de Macario, E. C.; Fahey, J. L.; Friedman, H.; Penn, G. M. Eds.; American Society of Microbiology: Washington D. C., 1992, pp. 678-684.
15. Bock, S. A.; Buckley, J.; Holst, A.; May, C. D. Clin. Allergy 1977, 7, 375-383.
16. Bock, S. A.; Sampson, H. A.; Atkins, F. M.; Zeiger, R. S.; Lehrer, S.; Sachs, M.; Bush, R. K.; Metcalfe, D. D. J. Allergy Clin. Immunol. 82, 986-997.
17. Taylor, S. L.; Lemanske, R. F., Jr.; Bush, R. K.; Busse, W. W. In *Food Allergy*, Chandra, R. K. Ed.; Nutrition Research Education Foundation: Newfoundland, 1987, pp. 21-44.
18. Harrison, L. A.; Bailey, M. R.; Naylor, M. W.; Ream, J. E.; Hammond, B. G.; Nida, D. L.; Burnette, B. L.; Nickson, T. E.; Mitsky, T. A.; Taylor, M. L.; Fuchs, R. L.; Padgette, S. R. J. Nutr. 1996, 126, 728-740.
19. Breiteneder, H.; Ebner, C. J. Allergy Clin. Immunol. 2000, 106, 27-36.

20. Food & Agriculture Organization. Report of a Joint FAO/WHO Expert Consultation on Allergenicity of Foods Derived from Biotechnology, Rome, Italy, Jan. 22-25, 2001, 27 pp.
21. Arshad, S. H.; Malmberg, E.; Kraft, K.; Hide, D. W. Clin. Exp. Allergy 1991, 21, 373-376.
22. Nordlee, J. A.; Taylor, S. L.; Townsend, J. A.; Thomas, L. A.; Bush, R. K. N. Engl. J. Med. 1996, 334, 688-692.

Chapter 17

Prediction of Parental Genetic Compatibility to Enhance Flavor Attributes of Peanuts

H. E. Pattee[1], T. G. Isleib[2], F. G. Giesbrecht[3], and Z. Cui[2]

[1]Market Quality and Handling Research Unit, Agricultural Research Service, U.S. Department of Agriculture, Campus Box 7625, North Carolina State University, Raleigh, NC 27695
[2]Department of Crop Science, Campus Box 7620, North Carolina State University, Raleigh, NC 27695
[3]Department of Statistics, Campus Box 8203, North Carolina State University, Raleigh, NC 27695

As future advances in transformation technology allow insertion of useful genes into a broader array of target genotypes, the choice of targets will become more important. Targets should be genotypes that will pass to their progeny other useful characteristics, such as sensory quality characteristics, while improving agronomic performance or pest resistance. This is particularly important if flavor quality is to be maintained or improved as the transgene is moved into breeding populations via sexual transfer. Selection of genotypes with superior breeding values through the use of Best Linear Unbiased Prediction procedures (BLUPs) is discussed and using a database of sensory attributes on 250 peanut cultivars and breeding lines, the application of BLUP procedures to the selection of parents for improvement of roasted peanut and sweet attributes in breeding of peanut cultivars is illustrated.

Currently, transformation can be used to insert useful genes into specific regenerable genotypes of many crop species. The transgenes are then moved into commercial cultivars by backcrossing. For example, in peanut (*Arachis hypogaea* L.), transformation mediated by *Agrobacterium tumefasciens* has been reproducible only in the obsolete cultivar New Mexico Valencia A (*1*). Transformation via microprojectile bombardment of somatic embryos is less genotype-specific, but the efficiency of regeneration of plants is highly dependent on genotype (*2*).

Future advances in transformation technology will permit insertion of useful genes into a broader array of target genotypes. With these advances, selection of target genotypes with superior quality traits and superior capacity to transmit those qualities to new cultivars will be more critical. Estimation of this capacity, termed "breeding value" in animal improvement and "combining ability" in plant improvement, is not a new concept. However, traditional methods of estimating breeding value require complex mating designs and extensive progeny testing. Therefore, selection of parents in conventional plant breeding is usually based on the individual's phenotype rather than on its breeding value. This short-cut method of parent selection can produce some inferior breeding populations.

Best Linear Unbiased Prediction (BLUP) is a procedure described by Henderson (*3*) to estimate the breeding values of dairy cattle based on data collected on all types of relatives rather than on progeny alone, obviating the need for complex mating designs and extensive progeny testing. Data on progeny of specific animals can be included in the analysis but are not required. The method is based on a mixed linear model with known variance-covariance structure among fixed and random effects. In general, the genetic effects in the model are considered to be random while the environmental effects are considered to be fixed. The variance-covariance matrix of additive genetic effects is calculated using standard quantitative genetic theory and is based upon the matrix of coancestries among related lines (*4*). BLUP is widely used in animal breeding and tree improvement (*5*) and is beginning to be used in annual crop species. Bernardo (*6, 7, 8, 9*) found it useful for identifying superior single crosses in maize (*Zea mays* L.) prior to field testing. Panter and Allen (*10, 11*) found BLUP to be superior to midparent value in selecting cross combinations in soybean (*Glycine max* L.).

Enhancement of roasted flavor of peanuts has been a long-standing objective of the peanut industry. Roasted peanut flavor has several attributes: roasted peanut, sweet, bitter, astringent, fruity, etc. and is the primary trait that induces consumers to buy peanuts. Highly significant correlations have been found among means for the attributes, particularly among roasted peanut, sweet and bitter (*12, 13*). The chemical basis of roasted peanut flavor is not well known, but is thought to be pyrizines derived from sugars and amino acids under heating. The specific genes or gene products involved in flavor precursor control are unknown. Through the research of Pattee and coworkers certain roasted peanut quality sensory attributes have been shown to be heritable (*13, 14, 15, 16, 17*). They have also shown that the choice of parents to create a new variety can influence flavor quality.

There are four market-types of peanuts, each with a different primary usage. The runner market-type is used to make peanut butter. Large-seeded virginia market-type are sold in-shell at ball parks, in grocery stores, and as boiled peanuts. They are also sold shelled as cocktail peanuts. Spanish market-type are used in confectionery products and mixed-nut products. Valencia market-type are sold in-shell in grocery stores. These market-types are genetically diverse in parentage and these differences can be important in selecting for breeding value. The runner and Virginia market-types have an alternate branching pattern typical of subspecies *hypogaea* and pod characteristics typical of botanical variety *hypogaea*. Their genetic base is predominantly the *hypogaea* botanical variety, but current cultivars and breeding lines have at least some ancestry from subspecies *fastigata*. The Spanish and Valencia market-types are entirely from the subspecies *fastigata* Waldron, the Spanish lines from botanical variety *vulgaris* Harz and the Valencia lines from botanical variety *fastigata*. Because the Virginia and runner market-types come from a distinctly different genetic background than the fastigate types, it is conceivable and perhaps likely that these differences can be important in sensory attribute relationships.

Our objectives are to (a) introduce Best Linear Unbiased Prediction procedures (BLUPs), which can help select the genotypes with superior breeding values, and (b) present the concept that parent selection becomes more critical as the capacity to insert transgenes into target genotypes improves because of the wider availability of genotypes.

Materials and Methods

Genotype Resources. The data used for this study were gathered over an 11-year span and include four peanut market-types, 250 different genotypes and 53 environments (year-by-location combinations). In the data set there are 1822 observations on roasted peanut attribute, 1779 on sweet and bitter attributes, and 1460 on the astringent attribute. All samples were obtained from plants grown and harvested under standard recommended procedures for the specific location. The market-types Spanish and Valencia have been combined in the data set because of an insufficient number of Valencia entries to properly represent the group.

Sample Handling. Across years samples were shipped to Raleigh, NC in February following harvest and placed in controlled storage at 5 °C and 60% RH until processed.

Sample Roasting and Preparation. The peanut samples were roasted between May and June using a Blue M "Power-O-Matic 60" laboratory oven, ground into a paste, and stored in glass jars at -10 °C until evaluated. The roasting, grinding, and color measurement protocols were as described by Pattee and Giesbrecht (*18*).

Sensory Evaluation. A long-standing six to eight-member highly-trained roasted peanut profile panel at the Food Science Department, North Carolina State University, Raleigh, NC, evaluated all peanut-paste samples using a 14-point intensity scale. Panel orientation and reference control were as described by Pattee and Giesbrecht (*18*) and Pattee *et al.* (*14*). Two sessions were conducted each week on nonconsecutive days.

Statistical Analysis. PROC MIXED in SAS (*19*) was used for analysis of the unbalanced data set to estimate the sensory attribute least square means for genotypes. Covariates fruity and roast color were used, as needed, based upon the findings of Pattee *et al.* (*12, 20, 21*). The fixed effects were genotype, region, genotype-by-region, and covariates fruity and roast color. Each genotype effect was partitioned to reflect the effects of market type and genotype within market-types. Classification of lines into market-types was based upon branching pattern, pod type, and seed size.

PROC IML in SAS was used to perform the calculations to compute BLUP estimates given in Harville (*22*). The mixed model (Formula 1) includes a parameter for the population mean (μ), a set of fixed effects (β) with a corresponding incidence matrix (X) that assoviates specific effects with individual observations, a set of random additive genetic effects (a) with its incidence matrix (Z), and a vector of error terms (ϵ):

$$Y = \mu + X\beta + Za + \epsilon \tag{1}$$

The variance-covariance matrix for the random effects and error terms is

$$\operatorname{Var}\left(\begin{bmatrix} a \\ \epsilon \end{bmatrix}\right) = \begin{bmatrix} G & 0 \\ 0 & R \end{bmatrix}\sigma^2 \tag{2}$$

where σ^2 is the error variance and $A = \operatorname{Var}([a]) = G\sigma^2$ is the additive genetic variance-covariance matrix for the lines. G is therefore $2Ch^2/(1-h^2)$ where C is the coancestry matrix and h^2 is the narrow-sense heritability of the trait. Pedigree information on the lines was obtained from published records and from the individual breeders. Coancestries among lines were calculated using standard computational techniques incorporated into the computer program of Delannay *et al.* (*23*). Modifications described by Cockerham (*24*) were required to calculate coancestries among lines derived from the same cross. Lines tracing to different F_2 plants had the same coancestry as full sibs, while pairs tracing to the same F_2 or later generation selection were more closely related than full sibs. When no information was available on the commonality of two lines derived from the same cross, it was assumed that the lines traced to different F_2 selections.

The standard BLUP solutions (Formula 3) can be obtained only when the genetic variance-covariance matrix is nonsingular.

$$\begin{bmatrix} \hat{\beta} \\ \hat{a} \end{bmatrix} = \begin{bmatrix} X'R^{-1}X & X'R^{-1}Z \\ Z'R^{-1}X & Z'R^{-1}Z+G^{-1} \end{bmatrix}^{-1} \begin{bmatrix} X'R^{-1}Y \\ Z'R^{-1}Y \end{bmatrix} \tag{3}$$

Because of the inclusion of multiline cultivars and their component pure lines in the study, there were collinearities in the coancestry matrix, the G matrix was singular, thus the variance-covariance matrix cannot be estimated for BLUPs calculated in this way (21). The BLUP solutions for a singular G matrix were obtained using Formula 4.

$$\begin{bmatrix} \hat{\beta} \\ \hat{v} \end{bmatrix} = \begin{bmatrix} X'R^{-1}X & X'R^{-1}Z \\ Z'R^{-1}X & Z'R^{-1}ZG+I \end{bmatrix}^{-1} \begin{bmatrix} X'R^{-1}Y \\ Z'R^{-1}Y \end{bmatrix} \text{ where } v = G^{-1}a \tag{4}$$

PROC CORR, PROC GLM and PROC GPLOT in SAS (19) were used to perform other statistical analyses in this chapter.

Results and Discussion

To better understand the impact of genetic variability on the sensory aspects of crop quality characteristics of a species it is essential to also understand the various environmental sources of variability. As previously stated the three primary sensory attributes that are heritable are roasted peanut, sweet, and bitter. Some aspects of the variation in these flavor components have been investigated, such as the effects of roast color and the attribute fruity (*20, 21*), genotype-by-environment (GxE) interaction on roasted peanut, sweet, and bitter (*12, 15*), ancestral effects on roasted peanut attribute (*17*), and high oleic acid content (*25*). These results have not previously been brought together in single review.

222

Sources of Variation Within Heritable Roasted Peanut Sensory Data.

Environmental factors are the predominate source of variability in roasted peanut and bitter attributes while genotype is the single most important factor in the sweet attribute (Figure 1). Of the environmental effects, year stands out as a source of variation. Differences between years generally reflect differences in temperature and rainfall, although the three peanut-producing regions of the US (Virginia-Carolina, Georgia-Florida-Alabama, and Texas-Oklahoma) are separated by sufficient distance that one would not expect consistent climatic effects across all three. However, year-by-region interaction was small for all three attributes. Genotype-by-environment interaction (GxE) effects were small in comparison with genotypic variation for the sweet and bitter attributes, but relatively large for roasted peanut, especially the interaction of genotypes with specific locations within years and production regions. Each attribute has a substantial amount of error variation, i.e., variation that was not attributable to any of the factors included in the statistical model, suggesting that additional factors influencing flavor could be identified in the future.

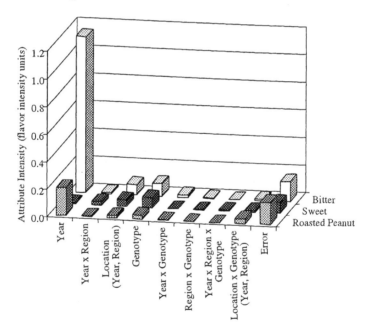

Figure 1. Magnitudes of variance components reflecting predominant sources of variation in flavor attributes of roasted peanuts.

The range of genotypic variation is different for the different market-types of peanut (Figure 2). The runner market-type has the greatest mean and the greatest maximum value for the roasted peanut attribute, followed by the fastigiate market-types and then by the Virginia market-type. However, the distributions of the three groups overlap. There is room to improve the roasted peanut scores of Virginia and fastigiate market-types, but there is also a risk of releasing runner cultivars with roasted peanut intensity inferior to Florunner, the long-time industry standard.

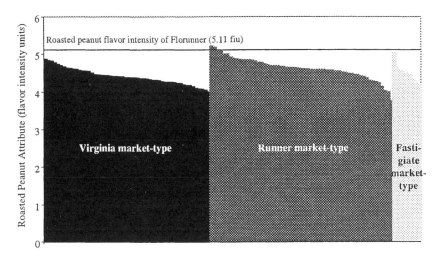

Figure 2. Means for roasted peanut attribute intensity across 122 peanut cultivars and breeding lines.

Substitution of Broad-sense for Narrow-sense Heritability in the G Matrix.

Because only broad-sense heritability (H) estimates are available for the sensory attributes (*13, 14, 16, 18*), BLUPs were computed for each sensory attribute using a range of estimates of narrow-sense heritability (h^2) (Table I, II, III). The estimates of h^2 bracketed the published estimates of H (0.06 to 0.11 for roasted peanut, 0.26 to 0.37 for sweet, and 0.02 to 0.06 for bitter). Because it reflects only the fraction of phenotypic variance caused by additive genetic effects, narrow-sense heritability must be less than or equal to broad-sense heritability which reflects all genetic variation. Correlations among BLUPs obtained using the heritability values were examined as indicators of the sensitivity of the technique to variation in the heritability estimate. In all cases, the correlations and rank correlations among the BLUPs were very high,

Table I. Correlations among BLUPs of breeding value for the roasted peanut attribute estimated at selected heritabilities.

Heritability estimate	Correlation		Rank correlation	
	$h^2=0.10$	$h^2=0.15$	$h^2=0.10$	$h^2=0.15$
$h^2=0.05$	0.9879	0.9690	0.9720	0.9454
$h^2=0.10$	--	0.9954	--	0.9933

Table II. Correlations among BLUPs of breeding value for the sweet attribute estimated at selected heritabilities.

Heritability estimate	Correlation		Rank correlation	
	$h^2=0.20$	$h^2=0.25$	$h^2=0.20$	$h^2=0.25$
h=0.15	0.9972	0.9911	0.9997	0.9980
h=0.20	--	0.9983	--	0.9980

Table III. Correlations among BLUPs of breeding value for the bitter attribute at selected heritabilities.

Heritability estimate	Correlation	Rank correlation
	$h^2=0.10$	$h^2=0.10$
$h^2=0.05$	0.9918	0.9908

indicating that the method is relatively insensitive to imprecision in the heritability estimate used in the calculations.

BLUPs of Breeding Value for Roasted Peanut and Sweet Attributes

Using a database of sensory attributes on 250 peanut cultivars and breeding lines, BLUP procedures were used to predict breeding values of parents for the roasted peanut and sweet attributes of peanut flavor (Figure 3). The range of predicted breeding values for roasted peanut attribute was -0.51 to +0.45 flavor intensity units (fiu), approximately twice the range of flavor intensity needed to establish a statistically significant difference. The range for sweet attribute was -0.65 to +0.68 fiu. The range for bitter was -0.41 to +0.40 fiu (data not shown). These values indicate that there is genetic potential to improve flavor quality through breeding. In collecting the sensory data, panelists assigned whole number scores to each sample, so these ranges are sufficiently large to be detectable by the human palate. The correlation observed between BLUPs for roasted peanut and sweet (r=0.71, P<0.01) was similar to the correlation

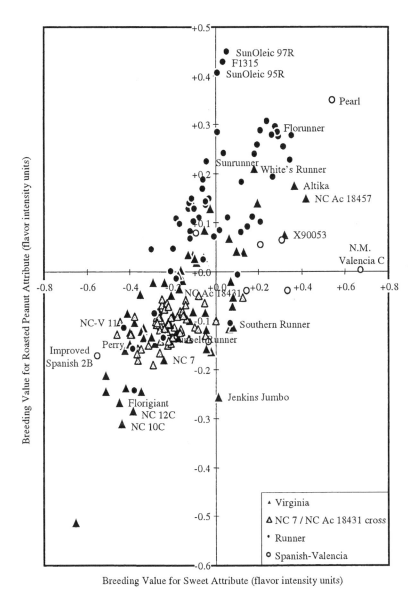

Figure 3. Estimated breeding values for roasted peanut and sweet sensory attributes in 231 cultivars and breeding lines.
SOURCE: Reproduced from Reference 26.

observed between genotypic means for the two traits (*13*). The biological cause of the correlation cannot be determined from this data. It could arise from genetic linkage or pleiotropism, or it could be an artifact of sensory perception.

Based on the BLUPs, several parents had superior predicted impacts on flavor quality. These included Florunner, its component lines and progeny. Because of its dominance as a cultivar over a 20-year period, Florunner was widely used as a parent. This has provided the array of runner-type cultivars descended from Florunner with advantageous flavor profiles. It has been documented that the runner population has superior average flavor profiles in comparison with the other market-types (*13, 16*). This superiority appears to extend to breeding values as well.

Several of the lines subjected to sensory evaluation were derived by backcrossing the recessive high-oleate character into the Sunrunner cultivar (*27*). These backcross derivatives included released cultivars SunOleic 95R and SunOleic 97R and some unreleased lines. All four high-oleate lines had substantially greater breeding values for roasted peanut attribute than did the recurrent parent Sunrunner. The high-oleate character is known to extend the shelf-life of peanut products by reducing the off-flavors caused by oxidation of linoleic acid (*287, 29*). It remains to be seen whether the impact of the high-oleate trait on the breeding value for roasted peanut attribute is due to a positive effect on the attribute itself or to the lessened presence of the painty and stale attributes associated with oxidation during storage of the samples.

The Spanish-type germplasm line Pearl had the best breeding value for roasted peanut attribute of any fastigiate line evaluated. New Mexico Valencia C had the highest breeding value for sweet attribute of any line, but it was only slightly higher than Pearl, and New Mexico Valencia C was neutral in its effect on roasted peanut.

The Virginia market-type has been shown to have lower average roasted peanut and sweet intensities and higher bitter intensity than the runner market-type. A genetic basis for this general inferiority may be inferred from the breeding values of certain key ancestors of the large-seeded virginia-type breeding population. Almost all large-seeded peanut lines trace their ancestry, at least in part, to Jenkins Jumbo, a line with very weak roasted peanut scores. It also has a strongly negative breeding value for roasted peanut. This line was a grandparent of the cultivar Florigiant which, like its relative Florunner in the runner market-type, dominated production in the Virginia market-type over a 15-year period. Florigiant was widely used as a parent in Virginia-type breeding programs, passing its poor flavor profile to its descendants. Another ancestor with a deleterious effect on flavor was Improved Spanish 2B, one of seven parents used to initiate the breeding program in North Carolina in the 1940s.

Several of the Virginia-type lines with the lowest predicted breeding values for roasted peanut attribute are lines resistant to Cylindrocladium black rot (CBR), a soil-borne disease caused by *Cylindrocladium parasiticum* Crous, Wingfield, & Alfenas. These include released cultivars NC 10C and NC 12C. The latest CBR-resistant cultivar to be released is Perry which has slightly better roasted peanut intensity than the other resistant cultivars but a substantially better predicted breeding value for roasted peanut.

Among Virginia-type lines, White's Runner, Altika, NC Ac 18457, and X90053 had superior profiles. White's Runner was another of the seven parents used to establish the North Carolina breeding program. It was a parent of NC 2 which has a relatively high roasted peanut attribute intensity for a commercial Virginia-type cultivar. Through NC 2, White's Runner is also an ancestor of Perry, the CBR-resistant line with the best roasted peanut intensity and breeding value. NC Ac 18457 was selected from a cross between Shulamith, an Israeli cultivar related to Florigiant, and Robut 33-1, a Spanish-type line from the International Crop Research Institute for the Semi-Arid Tropics (ICRISAT) in Hyderabad, India. Because of the incorporation of the unusual Spanish ancestry, NC Ac 18457 represents a somewhat different gene pool than the other Virginia-type lines in the population. Likewise, X90053 which derives its large-seed size from Japan Jumbo rather than from Jenkins Jumbo as do most other large-seeded Virginia-type lines, is a source of novel genes for seed size and perhaps for flavor attributes as well.

One hundred F_3-derived families from the cross of NC 7 and NC Ac 18431 were included in the flavor evaluation. This population provides the greatest opportunity to observe the distribution of flavor profiles arising from a single cross. For roasted peanut attribute, the two parents were separated by 0.15 fiu, and the breeding values of the segregates were all intermediate to those of the two parents, as expected with polygenic control of a trait. However, for the sweet attribute, there was transgressive segregation, and the predicted breeding values accordingly extend beyond the narrow range defined by the two parents (0.07 fiu).

The BLUP estimates provide a basis for choosing targets for transformation. There are eight lines whose breeding values make them obvious candidates for use as parents in a program of flavor improvement: runner-types Florunner and SunOleic 97R; Spanish-type Pearl, Valencia-type New Mexico Valencia C, and Virginia-types Altika, NC Ac 18457, White's Runner, and X90053. Most of these combine superior breeding values for roasted peanut and sweet attributes; SunOleic 97R is a superior parent for improvement of roasted peanut but neutral for sweet while the converse is true for New Mexico Valencia C. This array of genotypes includes representatives of all four US market-types and three botanical varieties. It should offer a range of responses to transformation and regeneration techniques.

Validation of BLUP Estimates

BLUP estimates were validated by calculating the correlation between predicted and observed values of sensory attributes for individual lines for whose ancestors breeding values were estimated (Table IV). Because the BLUPs would be used to predict the mean of hybrid populations, correlations were also calculated between predicted values and the means of lines derived from the same cross. Correlations of observed cross means with means predicted by BLUPs were above 0.85 for all three attributes. The correlations for individual lines were lower due to the variation to be expected among the progeny of a single cross. Based on these correlations, the BLUPs account for 74 to 84% of the variation in flavor attributes among cross means. This degree of correlation indicates a strong agreement between the observed cross means and the values predicted using BLUPs.

Table IV. Correlation of observed values with values predicted by BLUPs of breeding value or by midparent value

	Roasted Peanut	Sweet	Bitter
Observed value vs. value predicted by BLUP			
231 observations on lines [a]	0.6335	0.6520	0.7046
53 cross means	0.8839	0.9152	0.8588
Observed value vs. value predicted by BLUP or MP			
143 observations on lines, BLUP [b]	0.3325	0.2610	0.2675
143 observations on lines, MP	0.2729	0.1788	0.2470
13 cross means, BLUP	0.8220	0.7743	0.6189
13 cross means, MP	0.6803	0.6025	0.5771

[a] 231 lines from 53 crosses had BLUP estimates for both parents, permitting prediction of the cross mean from the BLUPs of the parents.
[b] 143 lines from 13 crosses had both parents or all four grandparents included in the genotype sample, permitting prediction of the cross mean by the midparent value.

In the absence of data from progeny tests, the traditional method of predicting the value of a cross mean is to calculate the mean of the two parents (10). This predictor is based on the assumption that most of the genetic variation for a trait is additive in nature. Correlations were calculated between observed values and midparent or mid-grandparental values to permit comparison between BLUPs and midparent values as predictors of line and cross performance. In each case, the BLUPs were at least as good as the midparent; for roasted peanut and sweet attributes, the correlation of the BLUP-based predictors with observed values of progeny were substantially higher than those for the mid-parent predictors, particularly for the cross means. In spite of the somewhat more complex calculations required to obtain the BLUP estimates, they appear to increase the precision of prediction of cross performance above that attainable using more traditional methods.

Conclusion

Best linear unbiased prediction of breeding value for flavor attributes was used to identify several peanut cultivars and breeding lines that would be likely candidates for transformation with agronomically useful genes. Once transformed, these lines would pass to their progeny not only the transgenes, but also superior flavor quality. Historically, flavor quality has not been a primary objective in peanut breeding as have agronomic traits such as yield and disease resistance. The result has been random variation in flavor quality of the different market-types. Heavy use of Florunner in the runner market-type and similar use of Florigiant in the Virginia market-type are the primary causes of the divergence in flavor quality between the two groups. The same type of divergence could result from widespread use of a parent selected solely on the basis of its ability to be transformed and regenerated. The combined usage of BLUP estimates for agronomic and quality traits will enable breeders and plant molecular biologists to work in concert in improving economic plant species.

References

1. Cheng, M.; Hsi, D. C. H.; Phillips, G. C. *Peanut Sci.*, **1994**, 21, 84-88.
2. Ozias-Akins, P.; Anderson, W. F.; Holbrook, C. C. *Plant Sci.*, **1992**, 83, 103-111.
3. Henderson, C. R. *Biometrics* **1975**, 31, 423-447.
4. Malécot, G. Les mathématiques de l'hérédité. Masson et Cie, Paris. **1948**.
5. White, T. L.; Hodge, G. R. *Theor. Appl. Genet.* **1989**, 76, 719-727
6. Bernado, R. *Crop Sci.* **1994**, 34, 20-25.
7. Bernado, R. *Crop Sci.* **1995**, 35, 141-147.
8. Bernado, R. *Crop Sci.***1996a**, 36, 862-866.
9. Bernado, R. *Crop Sci.* **1996b**, 36, 867-871.
10. Panter, D. M.; Allen, F. L. *Crop Sci.***1995a**, 35, 397-405.
11. Panter, D. M.; Allen, F. L. *Crop Sci.***1995b**, 35, 405-410.
12. Pattee, H. E.; Isleib, T. G.; Giesbrecht, F. G. *Peanut Sci.* **1997**, 24, 117-123.
13. Pattee, H. E.; Isleib, T. G.; Giesbrecht, F. G. *Peanut Sci.* **1998**, 25, 63-69.

14. Pattee, H. E.; Giesbrecht, F. G.; Mozingo, R. W. *Peanut Sci.* **1993**, 20, 24-26.
15. Pattee, H. E.; Isleib, T. G.; Giesbrecht, F. G. *Peanut Sci.* **1994**, 21, 94-99.
16. Pattee, H. E.; Giesbrecht, F. G.; Isleib, T. G. *Peanut Sci.* **1995**, 22, 158-162.
17. Isleib, T. G.; Pattee, H. E.; Giesbrecht, F. G. *Peanut Sci.* **1995**, 22, 42-48.
18. Pattee, H. E.; Giesbrecht, F. G. *Peanut Sci.* **1990**, 17, 109-112.
19. SAS Institute Inc. *SAS/STAT Software: Changes and enhancements through release 6.12*, SAS Inst., Inc, Cary, NC. **1997**.
20. Pattee, H. E.; Giesbrecht, F. G.; Young, C. T. *J. Agric. Food Chem.* **1991**, 39, 519-523.
21. Pattee, H. E.; Giesbrecht, F. G. *J. Sensory Studies.* **1994**, 9, 353-363.
22. Harville, D. A. *J. Amer. Stat. Assoc.* **1977**, 72, 320-340.
23. Delannay, S.; Rodgers, D. M.; Palmer, R. G. *Crop Sci.* **1983**, 23, 944-949.
24. Cockerham, C. C. *Crop Sci* **1983**, 23, 1177-1180.
25. Pattee, H. E.; Knauft, D. A. *Peanut Sci.* **1995**, 22, 26-29.
26. Pattee, H.E., Isleib, T.G., Gorbet, D.W., Giesbrecht, F.G., Cui, Z. *Peanut Sci.* **2001**, 28, __-__.
27. Gorbet, D. W.; Knauft, D. A. *Crop Sci.* **1997**, 37, 1392..
28. O'Keefe, S. F.; Wiley, V. A.; Knauft, D. A. *J. Am. Oil Chem. Soc.* **1993**, 70, 489-492.
29. Mugendi, J. B.; Sims, C. A.; Gorbet, D. W.; O'Keefe, S. F. *J. Am. Oil Chem. Soc.* **1998**, 75, 21-25.

Chapter 18

Outlook for Consumer Acceptance of Agricultural Biotechnology

D. B. Schmidt

International Food Information Council, 1100 Connecticut Avenue, NW, Suite 430, Washington, DC 20036

According to qualitative and quantitative research conducted in the US by the International Food Information Council (IFIC) consumers accept food biotechnology when its benefits are clearly understood. Consumer confidence in food biotechnology is dependent on food safety determinations *and* on effective communications. Government, academia, industry, and professional societies must work in partnership to help consumers and opinion leaders understand the benefits of biotechnology.

Consumer understanding and acceptance is crucial to the future of food biotechnology, and consumer views vary dramatically from country to country. Opinion in Europe has been consistently opposed to food biotechnology, yet acceptance remains fairly positive in the United States. We are at a critical point in agricultural biotechnology. How we address the issues over the coming months is going to be essential to how this technology proceeds over the next few years.

The long-term outlook for agricultural biotechnology is bright. It has been well said that "the genie is out of the bottle." Like many other technologies we have witnessed, biotechnology offers invaluable knowledge that cannot be stifled but will benefit mankind well into the future.

European-based activist groups experienced some success in 1999 in spreading fear and doubts about biotechnology in the US. They are unlikely to reach the levels of opposition they created in Europe and elsewhere unless government, industry, health professionals and academia here retreat from the science based principles of food regulation that have served us so well to date. It is not necessarily this opposition and resulting media coverage that threaten the future of agricultural biotechnology. Rather it is how regulators, industry, health groups and the university community respond to such activism that will determine the future. The American system of science and risk based food regulation has been put on the defensive by simplistic arguments based on uncertainty and social concerns. Universities are beginning to awaken as they watch years of careful research, in the form of field biotechnology experiments, and huge investments being destroyed in the middle of the night by the most radical of opposition groups.

Consumer research provides some clues on how we can navigate through these difficult issues. The International Food Information Council (IFIC) has conducted research since 1992 to understand the emotions consumers would go through when considering technology like food biotechnology. This research was done in extensive focus groups in ten United States cities. This was before any of the products produced from biotechnology had come to market. We wanted to understand the consumer and share this information so that we would know how to communicate with them when these issues surfaced down the road.

From this 1992 research we developed a schematic that we refer to as "the logic of emotion." This consists of five key points that scientific communications must touch on as consumers consider biotechnology: identity, scope, time, energy and benefits.

First is the *identity*, the whole idea that science and scientists can be good and bad. It is important for consumers to know the values of the people behind the technology. Concerns about *scope* indicate that some people look at this as "messing with Mother Nature" so it is important to communicate a scientist's respect for nature and the limits to which biotechnology will go.

It is easy to see in opinion polls that the support for human cloning, for instance, is almost nonexistent. When questions refer to animals, support increases but consumers are still split. As the subject changes to plant biotechnology, further support. is found

Time is a critical issue. It is important to tie biotechnology to the agricultural developments that have come before it, so that biotechnology is presented as an evolutionary, not a revolutionary, step. The idea of speeding something forward is not very reassuring to consumers. However, the idea that biotechnology is another set of tools that are consistent with what we've done for centuries – building on traditional agricultural practices – only now with more precision, is more reassuring.

It is easy as scientists and health professionals, with all the amazing breakthroughs that have taken place, to discuss biotechnology in high-tech, revolutionary terms. However, when communicating with consumers, this type of language is counterproductive.

Scientific jargon can be very confusing and alarming to consumers. Food biotechnology needs to be taken down to lay terms. It is important for consumers to understand that biotechnology is still about seeds that are planted in the ground, and they grow to be plants just like any other plants. If we don't speak at a consumer level, some of the language used sends consumers the message that we are talking about concoctions in laboratories. Consumers are not comfortable with that idea and it gives a false impression of what food biotechnology really is.

Energy is another important issue. In the United States, the American work ethic is very important. There has been a tremendous amount of research in the area of biotechnology. It is important to explain there hasn't been just one study in a laboratory one day, then an introduction to the food market the next day.

Most importantly, the *benefits* are what it is all about. It's "what's in it for me?" We have to explain up front why biotech products are being developed in the first place. Recently, the use of biotechnology has begun to provide benefits not only for growers, but also consumers. Growers reap higher crop yields, while consumers have greater product choices year round. More recently, media attention toward this subject has increased tremendously, and now may be a good time to take a look at this important advancement in agriculture and what benefits it holds for the future.

By producing crops that have been genetically enhanced, consumers can expect:

- Reduced levels of natural toxins in plants;
- Peanuts with improved protein balance;
- Food crops grown with fewer pesticide applications;
- Fungal resistant bananas;
- Tomatoes with a higher lycopene content;
- Fruits and vegetables enhanced for better nutrition and quality; and
- Increased crop yields and improved freshness and flavor

While some surveys have suggested that most Americans demand labeling of biotech foods, the IFIC surveys have been the only public vehicles to test consumer reaction to the actual Food and Drug Administration (FDA) labeling policy. The FDA requires special labeling when the use of biotechnology introduces a known allergen or when it substantially changes the food's nutritional content or its composition. Of consumers surveyed in May 2000, seven out of ten Americans (69 percent) support the current FDA labeling policy. In fact, most consumers felt that simply labeling a food as a product of biotechnology was not enough information to make an informed decision. This survey also found 86 percent of consumers agreed that it would be more useful for food manufacturers, government, health professionals and others to provide information on biotechnology through toll-free phone numbers, brochures and web sites, rather than changing the current FDA labeling policy (up from 81 percent in October 1999).

According to the May 2000 Wirthlin study conducted for IFIC, two out of three consumers support foods produced through biotechnology and have confidence in the FDA's policy for labeling biotech foods. The key to understanding American consumers' support for food biotechnology is the fact they trust their regulatory system. According to a March 2000 Gallup Survey, 80 percent of Americans have confidence in the FDA's regulation of the food supply. With such a high level of trust, fears about biotechnology have not become a major consumer concern as they have in Europe. It is the ultimate irony that food biotechnology has been rejected by the European Union in response to "public concerns", since the EU subsequently proposed to set up a food oversight authority similar to the US Food and Drug Administration (FDA) in order to bring European public confidence in food safety to the levels enjoyed by the FDA.

Qualitative research conducted with U.S. consumers in 1992, 1996 and 1998 found that consumers accept biotechnology, especially when the benefits are explained. In IFIC's May 2000 survey results showed 54 percent of consumers said they were likely to buy foods that have been enhanced to "taste better or fresher" (up from 51 percent in October 1999), and 69 percent if it had been modified for insect protection and to require less pesticide spray (up from 67 percent in October 1999).

How can that support be explained? Perhaps the answer lies in what the U.S. has *not* experienced. The U.S. has never experienced a crisis as frightening as BSE. And we do not have a history of food scares like those experienced in other parts of the world. Perhaps most important, the U.S. has watched the agencies responsible for managing such situations do so very effectively. Subsequently when FDA approved the use of products from food biotechnology U.S. consumers had every reason to believe they were safe.

We can't make 100 percent guarantees about the safety of anything, but there is an incredibly strong safety record with biotechnology. Given other technologies and food safety issues, biotechnology has an excellent record. We should be able to talk about that, but it is important to give an accurate rather than an absolute view of the safety of biotechnology. We cannot guarantee the future of anything, so we should not allow ourselves to be painted in a corner by some unforeseen event. DNA pioneer James Watson said it best recently: "To argue that you don't know what is going to occur is true about everything in life. People wouldn't get married, have children or do anything."

The vast majority of American consumers still place a great deal of confidence in the benefits of, and current regulatory climate for, agricultural biotechnology. A three-fold increase in media coverage of food biotechnology over the past year and confusion in the international marketplace have raised questions with some consumers. But most people remain positive and look forward to the benefits biotechnology will bring to the table. U.S. consumers are generally willing to accept biotechnology, especially when the benefits are explained.

When we discuss consumer response to biotechnology, we have found it is very important to look at how consumers put biotech in the continuum of food safety issues. Telephone surveys conducted in recent years by North Carolina State University have found that when consumers are asked, in an open-ended question, what food safety issues most concern them, biotech is rarely mentioned. This seems to hold true in Europe, as well. It might come up in the middle of the issues raised by Europeans, but in the U.S. it will rarely come up at all. In fact, most U.S. consumers focus on the common factors of taste, price and nutrition.

This suggests the importance of information sources. *Who* is educating consumers, not just the education itself, has emerged as a crucial factor to acceptance. Is it activists, government, academia? As we consider the impact of labeling and education on consumer acceptance of foods produced using agricultural biotechnology, the critical importance of language is often overlooked. How are they talking about this new technology and what language is being used?

Results of IFIC's qualitative and quantitative research provide insights on attaining optimal consumer acceptance of food biotechnology through the use of consumer-friendly terminology. Scientific jargon, although accurate, can invoke negative reactions from consumers unfamiliar with biotechnology. Therefore, mastering the language of food biotechnology has become a necessary discipline for all communicators of biotechnology.

We have made a real effort in the U.S. not to use phrases like "GMO" or "genetically altered." This is because these terms tap anxieties that precondition consumers to be wary of biotechnology, before they have even had a chance to understand the science and then make an informed decision about the products.

It is similarly important to explain what food biotechnology is. Focus groups reveal that consumers liken biotechnology to spraying the plants or "doing something to them" afterwards. Recent qualitative research conducted by Health Canada revealed that consumers believe the "genetically modified" products were never grown from seed in the first place.

In summary, the support for biotechnology demonstrated in the U.S. is largely based on situational differences. In many respects the future for the agrobiotechnology sector may be largely dependent on consumers ability to obtain information about this technology that is free of prejudicial language, balanced with cultural concerns.

Achieving that goal is pretty difficult. When you consider every party in this dialogue is committed to forwarding their particular opinion, and must use sensational language and aggressive tactics to harness attention, consumers rarely receive unbiased information about this topic. To address this issue, IFIC has developed ' The Communication Tenets for Consumer Acceptance of Agricultural Biotechnology', which is a ten-point reference that we urge all opinion leaders to use when communicating biotechnology issues to the public.

- First, the purpose of each new product of food biotechnology and its consumer benefits should be explained clearly at the beginning of every public discussion.
- Second, biotechnology should be placed in context with the evolution of agricultural practices.
- Third, emphasis should be placed on farmers who plant seeds that already contain beneficial traits developed through biotechnology.
- Fourth, an accurate rather than absolute, view of food and environmental safety determinations by regulators should be communicated for each product in each country.
- Fifth, communications should emphasize the research that led to the introduction of each new product of food biotechnology.
- Sixth -- this mostly applies to the U.S. -- communications should underscore that additional food labeling requirements are necessary only when there is a significant change in the composition, nutritional value or introduction of a potential food allergen from the gene transfer.
- Seventh, government and industry communications on food biotechnology must be consistent in order to earn consumer confidence.
- Eighth, it is important to distinguish between *consumer group activism* and *consumer attitudes*. Consumer group activism does not necessarily reflect consumer attitudes, and many consumer groups either support or do not oppose biotechnology.
- Ninth, multi-national approvals are the result of strong international scientific consensus.
- Finally, it is important to note that food biotechnology also provides important benefits in addressing hunger and food security throughout the world.

The International Food Information Council (IFIC) is a nonprofit organization founded in 1985 whose primary mission is to communicate science-based information on food safety and nutrition issues to the most influential opinion leaders for consumers, including health professionals, journalists, educators and government officials. IFIC is supported primarily by food, beverage and agricultural companies, and receives some funding from the U.S. government. IFIC does not play a role in lobbying or regulatory advocacy.

APPENDIX

International Food Information Council
"U.S. Consumer Attitudes Toward Food Biotechnology"

Wirthlin Group Quorum Surveys
October 8-12, 1999, February 5-8, 1999, March 21-24, 1997, and
May 5-9, 2000

Method

Approximately 1000 telephone interviews were conducted in March 1997, February 1999 and October 1999 among a national probability sample of adults 18 and older (stratified by state) in the continental United States. The range of error for a sample size of 1000 is +/- 3% at the 95% confidence level.

1. As you may know, some food products and medicines are being developed with the help of new scientific techniques. The general area is called "biotechnology" and includes tools such as genetic engineering. Biotechnology is also being used to improve crop plants. How much have you heard or read about biotechnology? Would you say you have read or heard . . . ?

	1997	Feb. 1999	Oct. 1999	May 2000
Total read or heard	79%	69%	73%	79%
A lot	11%	7%	13%	14%
Some	35%	26%	24%	31%
A little	32%	36%	36%	34%
Nothing at all	21%	31%	27%	21%
Don't know/refused	--	--	--	--

2. Now, using a 10-point scale, how well informed would you say you are about biotechnology, if **zero means you are not at all informed** about biotechnology and **ten means you are very well informed** about biotechnology

	Feb. 1999	Oct. 1999	May 2000
10	2%	2%	2%
9	0%	1%	1%
8	3%	5%	4%
7	4%	5%	6%
6	3%	5%	7%
5	9%	11%	14%
4	6%	11%	11%
3	16%	11%	12%
2	16%	13%	11%
1	39%	9%	11%
0	--	26%	21%
Don't know/refused	1%	1%	--

3. As far as you know, are there any foods produced through biotechnology in the supermarket now?

	1997	Feb. 1999	Oct. 1999	May 2000
Yes	40%	33%	38%	43%
No	37%	47%	38%	23%
Don't know/refused	23%	20%	24%	34%

3A. If yes, which foods? A total of 331 out of 1,000 participants responded. Total percentages are greater than 100% because multiple answers were given. **(New question for 1999 surveys).**

	Feb. 1999	Oct. 1999	May 2000
Vegetables	29%	42%	45%
Tomatoes	20%	27%	21%
Fruits	16%	23%	17%
Meats	16%	25%	16%
Produce/Processed foods	11%	5%	3%
Milk/Dairy	9%	10%	6%
Cereals/Grains	8%	6%	7%
Corn	6%	9%	18%
Lettuce	4%	1%	3%
Potatoes	3%	5%	3%
Soy	3%	3%	4%
Cheese	2%	--	--
Yogurt	2%	N/A	--
Strawberries	1%	1%	1%
Apples	1%	2%	2%
Grapes	1%	1%	1%
Melons	1%	1%	1%
Bananas	1%	--	1%
Eggs	2%	--	1%
Cucumbers	N/A	1%	--
Oranges	N/A	--	1%
Carrots	N/A	1%	1%

4.   All things being equal, how likely would you be to buy a variety of produce, like tomatoes or potatoes, if it had been modified by biotechnology to **taste better or fresher**?   Would you be very likely, somewhat likely, not too likely, or not at all likely to buy these items?

	1997	Feb. 1999	Oct. 1999	May 2000
Total Likely	**55%**	**62%**	**51%**	**54%**
Very Likely	19%	20%	18%	19%
Somewhat likely	36%	42%	33%	36%
Total Not Likely	**43%**	**37%**	**43%**	**43%**
Not too likely	21%	18%	18%	21%
Not at all likely	22%	19%	25%	22%
Don't know/refused	2%	1%	6%	2%

5.   All things being equal, how likely would you be to buy a variety of produce, like tomatoes or potatoes, if it had been modified by biotechnology to be **protected from insect damage and required fewer pesticide applications**? Would you be very likely, somewhat likely, not too likely, or not at all likely to buy these items?

	1997	Feb. 1999	Oct. 1999	May 2000
Total Likely	**77%**	**77%**	**67%**	**69%**
Very likely	39%	34%	28%	30%
Somewhat Likely	38%	43%	39%	39%
Total Not Likely	**23%**	**21%**	**27%**	**28%**
Not too likely	11%	11%	11%	14%
Not at all likely	12%	10%	16%	14%
Don't know/refused	1%	2%	6%	3%

6.   Biotechnology has also been used to enhance plants that yield foods like cooking oils. If cooking oil with reduced saturated fat made from these new plants was available, what effect would the use of biotechnology have on your decision to buy this cooking oil. Would this have a positive effect, a negative effect, or no effect on your purchase decision? **(New question for 1999)**

	Feb. 1999	Oct. 1999	May 2000
Positive effect	57%	42%	40%
Negative effect	10%	15%	18%
No effect	32%	39%	39%
Don't know/refused	1%	4%	3%

7. Do you feel that biotechnology **will provide benefits** for you or your family within the next five years?

	1997	Feb. 1999	Oct. 1999	May 2000
Yes	78%	75%	63%	59%
No	14%	15%	21%	25%
Don't know/refused	8%	10%	16%	16%

8. The U.S. Food and Drug Administration (FDA) requires special labeling when a food is produced under certain conditions: when biotechnology's use introduces an allergen or when it substantially changes the food's nutritional content, like vitamins or fat, or its composition. Otherwise special labeling is not required. Would you say that you support or oppose this policy of FDA?

	1997	Feb. 1999	Oct. 1999	May 2000
Total Support	**78%**	**78%**	**69%**	**69%**
Strongly support	45%	50%	45%	42%
Somewhat support	33%	28%	24%	27%
Total Oppose	**20%**	**19%**	**26%**	**28%**
Sonewhat oppose	9%	9%	12%	10%
Strongly oppose	11%	10%	14%	18%
Don't know/refused	2%	3%	5%	3%

9. Some critics of the U.S. FDA policy say that any food produced through biotechnology should be labeled even if the food has the same safety and nutritional content as other foods. However, others, including the FDA, believe such a labeling requirement has no scientific basis, and would be costly and confusing to consumers. Are you more likely to agree with the labeling position of the FDA or with its critics?

	1997	Feb. 1999	Oct. 1999	May 2000
FDA	57%	58%	50%	52%
Critics	40%	37%	45%	43%
Don't know/refused	3%	5%	5%	5%

10. Please tell me whether you: Strongly agree, somewhat agree, somewhat disagree, strongly disagree, don't know about the following statement.

Simply labeling products as containing biotech ingredients does not provide enough information for consumers. It would be better for food manufacturers, the government, health professionals and others to provide more details through toll-free phone numbers, brochures and web sites.
(New question October 1999)

	Oct. 1999	May 2000
Strongly agree	51%	55%
Somewhat agree	30%	31%
Somewhat disagree	7%	7%
Strongly disagree	5%	5%
Don't know/refused	7%	2%

INDEXES

Author Index

Subject Index

A

Acceptance of biotechnology. *See* Consumer acceptance of biotechnology
Acetyl coenzyme A, 120
Adjuvant, mucosal, 179
Advanced glycation end products (AGEs) in peanuts, 198–199, 199*f*, 201
Aflatoxins
  danger, in food, 132
  definition, 131
  insect damage, correlation with, 155, 157
  in peanuts, 152, 155
  production, 132, 152
Agrobacterium rhizogenes, 3
Agrobacterium tumefaciens
  pathogenic activities, 3
  transformation of alfalfa, 120–121
  tumor-inducing plasmid (Ti), 3
Alfalfa, genetically modified (GM), resveratrol glucoside in, 118–122, 120*f*, 126–127
Alkaloids, 43
Allelochemicals, 52, 61, 62
Allelopathic crops, risks and benefits, 63*t*
Allelopathy
  autotoxicity, 62
  comparison to synthetic herbicides, 61*t*
  definition, 52, 60
  genes, identification of, 61–62
  metabolic cost, 62
  risks and benefits, 63
Allergenicity of foods, 206

Allergens
  definition, 206
  in foods, 193
  induction of allergic sensitization, 207*f*
  sources in nature, 206
Allergic responses
  mediators, 206–208
  prevention, by induction of sensitization phase, 206, 207*f*
  tolerance, 182
Alternaria longipes, 107
Amadori products, in peanuts, 198–199, 199*f*
α-Amylase inhibitors, 43
2-Amino-4-(hydroxymethylphosphinyl)butanoic acid. *See* Glufosinate
AMPs. *See* Antimicrobial peptides (AMPs)
Anionic peroxidase, 157
Antibiosis, 38–39
Antibiotic resistance, 70, 73–74, 80–81
Antibodies, 172
Antifungal proteins
  general information, 77, 99, 100*t*–103*t*, 104–106
  genes from corn used in genetic engineering, 143
  See also β-1,3-Glucanases; Chitinases
Antigen presenting cells (APC), 172, 195
Antihistamines, 195–196
Antimicrobial peptides (AMPs)
  activity of peptides, 106–107

See also specific crops
Bt genes. See cry genes
Bt proteins
  characteristics of, 39–41
  expression in chloroplasts, 74–76,
    78–79
  in transgenic crops, 24, 26t
  resistance to, in insects, 31–33, 68,
    74–75
  toxicity, 25t
  See also Cry proteins
Bt resistance gene, 32
Butterflies, Monarch, 68, 75

C

Canadian National Farmers Union, 67
Canola, genetically modified (GM)
  amount planted, 1
  herbicide-resistant, 58
Cationic antibacterial peptides, 77–78
Chimeraplasty, 13
Chitinases
  antifungal activity
    in corn, 133–134, 140, 142f
    in transgenic plants, 100t
  effect on insects, 43
Chloroplast genetic engineering
  antibiotic resistance, 70, 73–74, 80
  betaine aldehyde dehydrogenase
    (BADH) gene from spinach, 80
  gene escape through pollen,
    prevention of, 67
  gene translation and transcription,
    69, 73
  genetic pollution, prevention of, 67
  herbicide resistance, 70–74
  insect resistance, 74–76
  limitations and challenges, 81–82
  marker-free, 70, 80–81
  maternal inheritance of genes, 68,
    72
  pathogen resistance, 76–78
  stress resistance, 69–70, 79–80
  tissue specificity, 68

  trehalose and drought tolerance,
    69–70, 79–80
Chloroplast transformations
  advantages of, 4, 79–80
  by particle bombardment, 73, 75
  difficulty of transforming, 41, 81–
    82
Cholera toxin (CT), 175
Chromosome number, doubling, 10
Codex Alimentarius Commission, 10
Cohen, Stan, 2–3
Colletotrichum destructivum
  (anthracnose), 104, 107–109
Colorado Potato Beetle (CPB),
  Leptinotarsa decemlineata, 18, 31,
  44, 47
Competition between crops and
  weeds, 59–60
Compositional comparisons of foods,
  206
Congeners, 67
Consumer acceptance of
  biotechnology
  activism, response to, 232
  benefits of biotechnology to
    consumers, 233, 234
  by global population, 5
  energy, as an issue in consumer
    acceptance, 233
  Europe, opposition in, 231, 232,
    234
  food safety issues, 234, 235
  identity, importance of, 232
  labeling policy, 233–234
  precautionary principle, 5, 11–12
  scope, concerns about, 232
  terminology, importance of, 235
  time, as an issue in consumer
    acceptance, 232–233
  United States, acceptance in, 231,
    232, 233–234
Corn
  aflatoxin in, 132–133
  antifungal proteins in, 133–144
  Aspergillus in, 132

RETURN TO: CHEMISTRY LIBRARY
100 Hildebrand Hall • 642-3753

LOAN PERIOD	1	2	3
4	**1-MONTH USE**	5	6

ALL BOOKS MAY BE RECALLED AFTER 7 DAYS.
Renewable by telephone.

DUE AS STAMPED BELOW		
~~NON-CIRCULATING~~ ~~UNTIL 10/31/03 BULL~~		

FORM NO. DD 10
3M 3-00
UNIVERSITY OF CALIFORNIA, BERKELEY
Berkeley, California 94720–6000